D1384543

INVESTING
IN
SOLAR STOCKS

INVESTING
IN
SOLAR STOCKS

An Investor's Guide to Winning in the Global Renewable Energy Market

JOSEPH BERWIND

New York   Chicago   San Francisco   Lisbon   London
Madrid   Mexico City   Milan   New Delhi
San Juan   Seoul   Singapore   Sydney   Toronto

1 2 3 4 5 6 7 8 9 0   DOC/DOC   0 1 0 9

ISBN   978-0-07-160895-4
MHID   0-07-160895-8

This publication is designed to provide accurate and authoritative information in regard to the subject matter covered. It is sold with the understanding that the publisher is not engaged in rendering legal, accounting, or other professional service. If legal advice or other expert assistance is required, the services of a competent professional person should be sought.
        —*From a Declaration of Principles Jointly Adopted by a Committee of the American Bar Association and a Committee of Publishers and Associations*

McGraw-Hill books are available at special quantity discounts to use as premiums and sales promotions, or for use in corporate training programs. To contact a representative, please visit the Contact Us pages at www .mhprofessional.com.

This book is printed on acid-free paper.

CONTENTS

ACKNOWLEDGMENTS

I would like to personally thank the many people who contributed to the success of writing *Investing in Solar Stocks*. I am very grateful to Gerard Visci, Ben Chinea, and Rick Fremont of Direct Access Partners in New York for providing me with support during the writing of this book. I am also grateful to Stephen Waldorf who came to the rescue when needed most, helping to manage the business when times were toughest; and to Dr. Subhendu Guha, Ph.D., Chief Technology Officer of Energy Conversion Devices, for his time and efforts; Christopher Tagatac, a 20-year veteran of Wall Street, who contributed much to the effort; and Dr. Aswath Damodaran, my finance processor from back in the day at New York University Stern School of Business, for his continuous and tireless efforts to broaden my understanding of the critical topics of valuation. It is also absolutely necessary to acknowledge the politicians of Germany and Japan, and naturally the German and Japanese people to whom the world owes a debt of gratitude, for supporting solar photovoltaic with well conceived financial incentives and support. In the USA, to Senator Bingaman and his team who I hold in high regard, thanks for taking the USA into a smart and well conceived policy to support solar energy until it is mature enough to stand alone. And finally, to my loving wife, Laura, and my daughters, Katherine and Elizabeth, who together with Daisy provide the sunshine in my life—this book is dedicated to you.

Introduction

During the 2008 presidential election the subject of alternative energy became one of the most hotly contested policy debates of the campaign, with both candidates seeking to claim leadership on the issue. Proposals were presented with much fanfare, and great promises were made on each side.

When politicians—the most cautious breed of animal known to the human race—consider an issue enough of a surefire winner to endorse it, it's an unmistakable signal that it has become a mainstream concern. In the years ahead we may look back on 2008 as the turning point for alternative energy, the moment that it outgrew its status as "alternative" and started to become the standard for the twenty-first century.

Solar stock investing is in its early days, and there has been no place to turn to for advice on just how to invest wisely. The principles behind successful investing in solar stocks remain shrouded in mystery. There are only a few institutional investment brokers with dedicated teams focused on solar energy, while many of Wall Street's heavyweights remain mired in a debate about whether the factors that create the alternative energy investment opportunity are here to stay. Meanwhile, private equity and hedge fund heavyweights are themselves getting into solar in classic fashion: early and big.

1

There are myriad reasons why alternative energy is an imperative for the future. Foremost among them, of course, is climate change and the environment. While there are still a few who do not agree that global warming is an imminent threat, the consensus is that clean, sustainable forms of energy are critical to protect the environment. With the energy needs of developing countries such as China and India growing at a rapid rate, pollution from fossil fuels has become a pressing concern.

National security is another impetus driving the adoption of alternative energies. The United States is dangerously reliant on foreign sources for oil. This became abundantly clear during 2007 to 2008, when oil prices skyrocketed to unprecedented levels, wreaking havoc on the economy. Prices have fallen back to earth since then, but it's plainly evident that in the long term oil will grow ever scarcer and ever more expensive. Add to that the fact that many of the countries from which we acquire oil are unstable at best and outright hostile at worst, and alternative energy clearly becomes a vital national security interest.

Yet while the environment and energy security loom over the discussion, in many ways they were only of peripheral importance during the 2008 presidential campaign. Indeed, most of the debate focused not on the environment or on energy security, as in previous years. Instead, proposals centered on the vast potential that alternative energy holds to stimulate economic growth, including 5 million new "green collar" jobs and $150 billion invested in alternative energy over the next 10 years. Of these, solar energy remains the largest source of potential clean and renewable energy, establishing it as the preeminent investment opportunity of our lifetime.

For solar stocks, this enthusiasm for alternative energy means that their day in the sun has arrived. It has been a long time coming, and informed investors will acknowledge that solar technology has not yet crossed the finish line. A brief history is in order.

FROM CELL TO SOLAR PHOTOVOLTAIC ENERGY

In 1839, Edmund Becquerel, a French physicist, discovered the photovoltaic effect while experimenting with an electrolytic cell made up of two metal electrodes. The photovoltaic effect is the basic physical process through which a solar cell converts sunlight into electricity.

Fast-forward to the 1950s when Bell Labs produced a silicon solar cell with 6 percent efficiency. By 1958 the U.S. ambition to have the first astronaut walk on the moon created funding for solar energy linked to the space program, where its first significant use was to provide power for the Vanguard satellite.

Since the oil shock in the early 1970s, when American dependence on foreign oil first received attention, the solar market has grown on average 25 to 30 percent per year. Nearly 40 years of market growth have been driven by government support, dramatic declines in the cost per watt ($/watt), technological advancements, manufacturing improvements, and sharp increases in capital flows. Broadly speaking, the cost of solar power has fallen by about 20 percent each time production doubled, which equates to annual price declines of about 5 percent a year.

While the worldwide electricity market is over a trillion dollars and growing, solar photovoltaic (PV) currently represents less than 0.5 percent of electricity generation globally. However, solar energy has been growing by 48 percent per year from 1996–2007, with more explosive growth expected through the end of the next decade, according to Alternative Energy Investing.™ As depicted in the chart in Figure 0.1, with 2003 as the base year, the energy technology sector has outperformed benchmark indices such as the Nasdaq Composite, S&P 500, and Russell 2000.

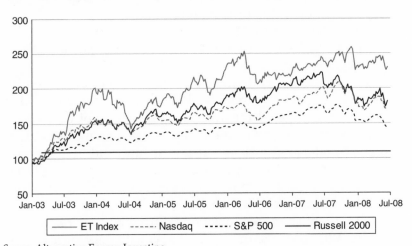

Source: Alternative Energy Investing

Figure 0.1 Performance of energy technology Index compared with benchmark indices

AN INVESTMENT OPPORTUNITY

The investment opportunity for solar stock investing is staggering. The solar industry is in the nascent stages of growth and has a long runway before it becomes a significant portion of overall energy production. For example, if the U.S. solar industry grows at 20 to 25 percent per year for the next 20 years, it would still account for less than 1 percent of total energy production. The prospect of high secular growth attracts investors seeking visibility and has driven large investments in solar companies throughout the value chain, leading to scale and scope benefits and reduced costs.

It should be noted that while solar photovoltaics offer tremendous investment opportunities, they are in fact commodities. However, the growing popularity of solar photovoltaic reflects specific attributes not enjoyed by other renewable energy alternatives today. Solar rooftop installations do not create the public opposition associated with wind farms visible on the horizon or with industrial-scale biomass power plants. Additionally, peak periods of solar photovoltaic output occur during peak electricity demand, so solar can provide peak power at the point of consumption, and therefore solar can reduce stresses on utility networks during periods of high demand. And finally, solar photovoltaic is frequently the low-cost power option for remote, off-grid locations, and although this market segment is small, it represents a growth opportunity as developing countries deploy solar photovoltaic.

Despite the enormity of the investment opportunity, solar has gone through up and down periods, and investing in it has not been easy. From late 2005 through the middle of 2008, prices for solar photovoltaic modules and systems actually rose as pressure from accelerating demand from both the PV and electronic semiconductor industries competed for insufficient supplies of polysilicon. While sales and margins rose throughout this period, producers acted fast to add to global silicon production, and by the end of 2008 multiyear silicon supply contracts (i.e., take-or-pay supply and sales contracts including upfront deposits), as well as module sales contracts, may have become an albatross to companies engaged in them to establish market positions during a temporary period of scarcity. Why did take-or-pay contracts with very large deposit payments for multiyear periods (i.e., 10 years or more) occur?

The notion that the solar industry would not face protracted periods of oversupply combined with a strategy to finance capacity growth backed by these supply contracts fostered a race to "lock-up" the supply chain. This became a leading method to raise capital, using what in retrospect appears to have been a toxic strategy.

At least two complications exist that negate the notion that oversupply is impossible. First, solar projects require relatively skilled design and relatively trained installation labor. The logic that Germany can act as a "backstop" market implies that the labor force can expand sufficiently to absorb any and all new solar modules entering the German market. Second, capital will exist at rates sufficient to finance projects at acceptable hurdle rates. In both cases, the year 2008 proved the case that oversupply can and indeed does emerge from time to time, resulting in vastly lower selling prices and breakeven to loss-making margins. As surplus modules flooded the German market from Spain and elsewhere, macroeconomic factors made the situation even worse, as capital dried up and interest rates used to calculate project returns in Germany caused many pending projects to get delayed or canceled altogether.

In the longer run, the question of elasticity of demand that affects solar sales and margins must be estimated with an eye to the elasticity of supply. As new supplies of materials (such as polysilicon) enter the market and flow from new investment, or slowdowns in adjacent markets (like semiconductors), or substitutes like upgraded metallurgical silicon, investors must normalize sales and earnings to reflect the more complicated markets characterized by government subsidies, at least until the industry achieves grid parity.

Estimators and consultants that led their colleagues and clients into believing that unlimited German subsidies meant that margins were ultimately sustainable are partially to blame in the unnecessary overinvestment in the solar industry that took place during the shortage period when companies reaped 100 percent of the subsidy benefits and solar stocks seemed to levitate to incredible new heights. These stocks came crashing down as new supplies flooded the market, and the macroeconomics of the credit crisis revealed the true nature of the cyclicality that continues to underpin the earnings drivers of the solar industry.

Some Key Questions

Key questions investors must first ask is whether installed costs for solar photovoltaic systems will decline appreciably because the silicon supply issues are resolved, and whether demand can sustain a tsunami of new module supplies from flooding the industry for years to come.

Second, should a glut of modules crash the market as appears to be the case in 2009, taking with it sales and margins throughout the supply chain, will public policy in the key market of Germany be sustained and can companies reduce costs by shedding boom-time supply contracts struck at high prices?

At the moment, half of the world market for solar photovoltaic systems resides in Germany, a country that has no limit, or "cap," on the volume of systems covered under the program. Thus, it is the world's market of last resort, and solar photovoltaic investments would certainly feel the pressure of a cap in the German market.

As support policies grow in the United States and elsewhere around the world, the investor must be cautious of policy reversals (i.e., U.S. state policies, Spain, and so on). Just as subsidy supports have grown, they may just as easily fall, especially as the credit crisis spreads around the world, putting exigent issues of civil welfare ahead of recently greatly reduced prices for fossil fuel energy.

And finally, will installed costs decline because of significant public sector investments to develop and grow local markets? If they don't, support programs worldwide could be curtailed. This would present a serious challenge to the industry because consistent, predictable government support has been required by the financial community. Therefore, it is crucial for the solar photovoltaic industry to expand sufficiently to balance supply and demand and allow price declines, thereby encouraging further market growth while mitigating the damage of boom and bust cycles.

ABOUT THIS BOOK

Investing in Solar Stocks is the essential guide to understanding one of the biggest investment opportunities of the twenty-first century. Written with an eye toward presenting concrete strategies for investors to profit from the explosion of solar technology, this book will advance the reader far beyond the competition by explaining clearly and in plain English the investment principles for buying, selling, and shorting solar stocks, including some helpful hints on successful investment strategies.

Don't let the tables and charts in the book fool you! Learning how to invest in solar stocks is not complicated. Successful investing is based on disciplined principles, and investing in solar stocks is no different. Investors will discover the principles to invest successfully from the ground up, starting with an overview in Chapter 1 of the technologies that are driving the solar industry today and those that may form the next big opportunity in the near future.

In Chapter 2 we outline the big picture, revealing how solar has grown into a global industry and discussing how local government policies play out across an international stage. A tour of the "solar system," in Chapter 3, breaks down the market forces and business models that shape the industry and introduce the solar supply chain. In Chapter 4 we delve into some of the valuation methods investors may use to unlock the profit-making potential of solar stocks. Chapter 5 reviews the risk and volatility factors that move solar stock prices and is focused in particular on how short-term investors can use these factors to their advantage. In Chapter 6 we continue the discussion of risk and volatility factors, this time from the perspective of long-term investors. Chapters 7 and 8 guide investors through diversification and rebalancing strategies for solar portfolios. And finally, in Chapter 9 we examine how the difficult economic situation that struck global markets in 2008 has affected the solar industry and what the long-term implications for solar stocks might be.

CHAPTER 1

Solar Technologies: The Science behind the Stocks

One of the biggest hurdles any new investor in solar stocks faces is getting up to speed on the technology. Solar technology is developing at an incredible pace, with new and exciting breakthroughs on the horizon, and investors stand to reap big rewards. However, as with all technologies, the rapid pace of progress can leave investors uncertain as to where to place their bets.

How can you tell which technologies are the most promising? It's easier than it looks. There are three important criteria to keep in mind when evaluating the technology of solar companies. Ranked in order of importance, they are:

1. *Cost is everything.* For almost any solar technology, the most important metric is electricity cost per kilowatt hour from the installation. Think "the cost of electricity at the wall outlet." Cost per kilowatt hour is the measure we use to pay our electric bills, and it indicates how competitive a particular solar technology is in producing energy compared to conventional electrical generation methods such as coal, gas, and nuclear power. It is difficult to establish industrywide benchmarks due to a highly complex utilities market in which the cost per kilowatt hour in one region may be much higher or lower than in another region. But the bottom line for the entire solar technology industry is

that companies are in a race to determine which company can produce electricity at the cheapest rate. This is the number that tells you who's winning that race.

- The critical question to ask: *What is the electricity cost per kilowatt hour from the installation?*

2. *Manufacturing matters.* As with any industry in the manufacturing sector, a company's survival ultimately depends on its ability to produce at high volumes with high yields. This is a significant challenge in the solar technology industry where "cost per watt" remains high because manufacturing processes are still being perfected and are subject to considerable defect rates. For instance, polycrystalline silicon—currently the dominant technology in solar and one we'll discuss at length later in the chapter—has a yield rate of only 95.5 percent, meaning almost 5 percent of products manufactured using this technology are lost to defects. There is much room for improvement in this area. Another technology we'll discuss, thin film, promises breakthroughs in efficient manufacturing.

- The critical questions to ask: *What are the current production yields, where do they need to be, how will they get there, and what changes to the production process are needed to achieve high volumes at high yields?*

3. *Conversion efficiency matters (sort of).* This term describes how efficiently solar cells convert sunlight into energy. Conversion efficiency can impact cost per watt either positively or negatively depending on many factors (i.e., raw material quality and cost, cell structure, etc.). Although higher conversion efficiency is usually better, conversion efficiency in and of itself is not the critical metric for the simple reason that the more efficient the cell, the more expensive it can be to produce. In fact, the highest-efficiency cells commercially available are extremely expensive. This defeats the purpose of solar technology: producing cheap energy.

- The critical questions to ask: *What are the current conversion efficiencies; where are they expected to go; when, how, and at what cost?*

These three criteria will allow any investor to understand the strength and weaknesses of the two major solar technologies in the marketplace, which we will now examine in greater detail.

SOLAR TECHNOLOGIES: SILICON VS. THIN FILM

Solar technologies can be split in to roughly two camps. Silicon-based technology, which currently accounts for more than 90 percent of the market for solar modules (Alternative Energy Investing™; company reports, 2008), is composed of the largest companies producing modules whose starting raw material is polysilicon. Its main competition is "thin film" technology, a relatively new process to the commercial scene that holds great promise for greater economies of scale and potentially lower costs (see Figure 1.1). Thin film currently accounts for 9.99 percent of the market, but this share is expected to grow to 15 to 20 percent by 2010 (Alternative Energy Investing™). The rest of the market is made up of various types, all in the very early stage and not destined for public company status anytime soon.

Within those two camps, solar technology is further differentiated by the specific kinds of materials and processes employed. In the silicon-based technology camp, there are primarily two different processes, polycrystalline production and monocrystalline production, with the polycrystalline production camp further subdivided into companies that produce either polycrystalline blocks, ribbons

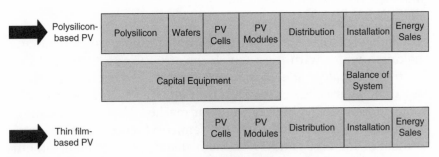

Figure 1.1 Solar PV value chain (crystalline silicon and thin film)

2008 Estimated MW Cell Production Mix

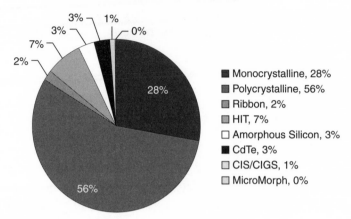

Monocrystalline, 28%
Polycrystalline, 56%
Ribbon, 2%
HIT, 7%
Amorphous Silicon, 3%
CdTe, 3%
CIS/CIGS, 1%
MicroMorph, 0%

Source: Alternative Energy Investing

Figure 1.2 2008 estimated MW cell production mix

or tubes. Within the thin film camp, there are several different processes associated with the specific materials each uses: amorphous silicon (a-Si), a-Si (μm)—microcrystalline silicon/amorphous silicon ("micromorph"), copper indium (gallium) selenide (CIS/CIGS), and cadmium telluride (CdTe).

The pie chart in Figure 1.2 shows how the overall solar marketplace is currently structured.

Drilling down into the details of these manufacturing processes is essential to understanding where the market has come from and where it is headed.

MANUFACTURING PROCESS FOR SILICON-BASED SOLAR

Silicon-based solar starts with the idea of taking polysilicon and making a wafer with it on which the structure to turn light into electricity can be made. Standard commoditized module products based on solar cells generating electricity at 15 to 17 percent efficiency dominate the bulk of the solar market. The core technology for silicon PV is derived from the semiconductor industry development of the 1960s and 1970s, with steady improvements in efficiency and manufacturing.

In this section, we explain the essentials of the manufacturing process for silicon-based photovoltaics. We start with a brief discussion on polysilicon before moving to ingots, wafers, cells, and modules. For cell manufacturing, we use a detailed example of a cell manufacturing line at Q-Cells based on a report issued by Alternative Energy Investing™ after factory visits to profile the line. Keep in mind that there are investment opportunities in public companies throughout the solar supply chain.

Polysilicon Manufacturing

A key point investors must remember when studying solar technology is that a solar cell is a kind of semiconductor:

> A *semiconductor* is a solid material that has electrical conductivity in between a conductor and an insulator. Silicon is used to create most semiconductors commercially, but dozens of other materials are used.
>
> —*Wikipedia*

This brief definition can be combined with the more specific definition to arrive at a basic understanding of exactly what a solar cell is without taking a physics class.

> A *solar cell* or *photovoltaic cell* is an electronic device that converts solar energy into electricity by the photovoltaic effect.
>
> —*Wikipedia*

While solar cells can be made from various compounds, including silicon, cadmium telluride, copper indium diselenide, and so on, the easily transferable developments in silicon technology from the electronics industry, and the long history of silicon-based semiconductors, make polysilicon by far the most popular material for solar cells. While the technology transfer from the semiconductor industry to the solar industry, and its popularity, seems to imply an ever-growing demand for polysilicon, the investor must beware of risks stemming from new entrants and the boom/bust cycle inherent in polysilicon.

The polysilicon industry has supplied the semiconductor industry with material of increasing purity for over 40 years, and while material quality and timely delivery have seldom been an issue, the profitability of polysilicon companies has been questionable at best. Investment risks throughout the solar supply chain can emanate from breathtaking price drops that typically occur in the polysilicon industry soon after new plants are constructed, resulting in over-supply until demand returns and once again outstrips supply.

Silicon (not polysilicon) is the second most abundant element in the earth's crust. In nature, it is found as an oxide (in the form of sand and quartz) and as a silicate (in the form of granite, clay, and mica). Silicon must be processed and purified into polysilicon to be used in the semiconductor or solar manufacturing process. This is where the distinction of semigrade, which is used for components in consumer electronics among other devices, and solar grade, which is used to make solar cells, comes in. Semigrade must be by far more pure than solar grade. Solar grade needs to achieve purity levels of only 99.9999 percent (often referred to as "six nines" pure), while semi-grade must be purified to nine nines at the low end and eleven nines at the high end. Therefore, although silicon is seemingly everywhere and always in abundant supply, polysilicon—the purified product of chemical companies like Hemlock (Dow Corning), Renewable Energy Corporation, MEMC, Tokuyama, and others—is a cyclical commodity, which is to say that it exhibits large price fluctuations.

As polysilicon companies vie for market share, they must pro-duce semiconductor or solar grades separately for each market. Often, polysilicon companies can achieve solar grade by producing at faster rates than the rates needed to achieve semigrade purity. Some polysilicon companies can run their processes up to 33 percent faster when producing solar grade, and if the price for solar grade is high, selling into the solar industry can become attractive even com-pared to selling semigrade, even though semigrade sells for higher prices due to its higher purity. This is due in part to the more effi-cient asset utilization when producing solar-grade polysilicon.

The polysilicon industry, using the standard Siemens tech-nology, requires a significant amount of capital and large amounts of electricity. It costs approximately $100 per kilogram ($100/kg) to build a standard polysilicon facility. For comparison purposes, a plant capable of producing 2,500 metric tons (MT) of polysili-con would cost $250 million to build. Such a plant could produce enough silicon for 275 megawatts (MW) of annual PV cell production at today's average silicon efficiency of nine grams per watt of PV.

The depreciation of this high capital cost along with variable manufacturing costs results in polysilicon costing approximately $0.60 per watt, or 18 percent of the total cost of today's polysilicon-based solar modules ($3.25 per watt). If the industry wants to bring average costs for modules down to $1.50 per watt, a level widely thought to be economically transformative in global markets, additional cost reductions in silicon processing will be required, but as one can see, not all the cost savings can be achieved by lowering polysilicon prices.

Understanding the changes in polysilicon technology is critical to this segment of the supply chain and to forming opinions about emerging technologies that have the potential to reduce the cost of producing solar modules and put the polysilicon-based solar companies at a competitive disadvantage. Moreover, higher cost polysilicon technologies may hamstring those companies whose supply-chain commitments are based on long-term contracts, when in the future other, lower cost polysilicon technologies and thin film solar could achieve lower costs and competitive advantage.

Polysilicon production is a capital intensive and highly technical business that requires large amounts of electricity for melting and purification. Currently, the United States is the global leader in polysilicon production. Hemlock, a subsidiary of Dow Corning in joint venture with a Japanese partner, remains the world leader. MEMC (ticker: WFR) has production in Texas using a novel production process. Renewable Energy Corporation (ticker: REC NO) also has production in the United States. But America's lead in polysilicon production as a function of cheap electricity and skilled labor is potentially giving way as European, Chinese, and Korean companies flush with investment funds, government subsidized electricity, and swiftly developing pools of engineers and skilled labor are building new polysilicon plants challenging U.S. dominance.

The investment implications of this rapid build-out is the risk of overcapacity and a glut causing contracts suffering from higher price levels to undergo considerable stress. Should oversupply of polysilicon lead raw material contracts to be renegotiated or broken outright, it could also lead to dislocation throughout the entire solar supply chain, upsetting sales and earnings estimates not only for polysilicon producers, but also for companies throughout the solar supply chain who have their own long-term contracts based on yesterday's prices.

The most common process for manufacturing solar-grade polysilicon is the "Siemens process," named after the company that first established the process. The raw material for the Siemens process is metallurgical-grade (MG-Si) silicon, which is essentially insufficiently pure silicon. In the Siemens process, hydrochloric acid (HCl) is reacted with MG-Si to form "trichlorosilane," or TCS. This TCS is distilled, and through a vapor deposition process using hydrogen, gets reduced to silicon. Polysilicon produced via this process is usually sold in the form of chunks, and is thus called "chunk polysilicon." The Siemens process has been in existence for several decades and is widely used in the industry. The advantages of building capacity based on the traditional Siemens reactors include the transparency and low risk of using a well-established and well-understood process. The Siemens process was used to produce approximately 90 percent of polysilicon in 2008.

For producers and new entrants into the polysilicon industry, using the Siemens technology is less risky, and comparatively easier to build, than other technologies, despite lower costs for those technologies. Siemens is the industry standard, but a few companies are involved with newer technologies that have been researched for decades and not yet brought to commercial scale production. These new technologies are more focused on a solar grade, or lower quality and thus cheaper, silicon product. This is achieved by modifying and simplifying the refining process.

One such newer technology is known as fluidized bed reactor technology, or FBR for short. This technology was developed by Ethyl Corporation and produces a highly refined final product of granular silicon. The technology starts with silicon fluoride instead of MG-Si, which is converted to monosilane, and then the silicon seed is dropped into the reactor while silane and hydrogen gases pass through. The process employs a lower capital cost and uses less electricity than Siemens reactors. To date, however, only MEMC—using a different raw material, fluorosilic acid—and Renewable Energy Corporation, have successful FBR plants, so the production economics are still unclear. Current estimates for FBR granule polysilicon is believed to be 15 to 20 percent cheaper than comparative fully depreciated polysilicon produced in a Siemens reactor (Table 1.1).

TABLE 1.1

Polysilicon Production Cost Based on FBR Technology

Polysilicon	FBR Granule Polysilicon
Plant scale	2,100 MT
Production cost	$42 million
$/kgs of polysilicon	$20.00
Source: Alternative Energy Investing	

Despite the cost advantages of FBR over Siemens technology, a new crop of polysilicon technologies are popping up to produce a cheaper alternative. These new entrants are building polysilicon plants exclusively for a solar industry that targets costs at half what today's technologies can achieve.

Upgraded Metallurgical Polysilicon

As new companies enter the market producing silicon specifically for the solar industry, new efforts to simplify the purification process are creating investment opportunities as well as risks. Upgrading the metallurgical silicon process could be a cost effective way to produce silicon for the PV industry. New entrants using upgraded metallurgical silicon technology could put substantially more cyclical pressure on the polysilicon industry as new supplies of "just barely" pure or "pure enough for solar" cheaper polysilicon hits the market.

A clarification on terminology is in order here. As discussed earlier, metallurgical grade silicon (MG-Si) such as the kind used in the Siemens process is the precursor to polysilicon. Upgraded metallurgical silicon bypasses the Siemens process, instead purifying metallurgical grade silicon through a series of steps. The upgraded material can then be added or "blended" with purer silicon to produce a polysilicon that, while not as pure or as high quality as the polysilicon produced by the Siemens process, is cheaper to manufacture. Currently available from Elkem, Dow (ticker: DOW), and Timminco (ticker: TIM CN), upgraded metallurgical silicon is currently being used in the market to produce 10 percent blends and up to 15 percent blends of upgraded metallurgical silicon to virgin material. Tests using rates between 50 and 100 percent are also being undertaken in a number of cell manufacturing labs, as we will shortly discuss.

The following list gives the purity levels of silicon produced by different processes:

- Standard metallurgical silicon MG-Si = one to two nines
- Siemens process (poly rods) = seven to eleven nines
- Fluidized bed reactor (poly beads) = seven to eight nines
- Upgraded metallurgical silicon = four to five nines

Upgraded metallurgical silicon's purity level is somewhere between that of standard metallurgical silicon and Siemens process-based polysilicon.

The advantage of upgraded metallurgical silicon is that it involves less capital costs and can be ramped up faster than the standard Siemens process. Combined with the fact that running a UMG-Si plant requires less energy, this could mean better economics compared to Siemens process-based plants.

Solar-grade silicon production using the upgraded metallurgical silicon method costs $15/kg versus $25 to $40/kg for Siemens production and $20 to $25/kg for FBR production. Renewable Energy Corporation forecasts that upgraded metallurgical silicon production could amount to 55,000 tonnes by 2011, with typical cell efficiencies of 15 percent. There are currently 15 manufacturing plants pursuing upgraded metallurgical solar silicon worldwide, with only five ready for large scale manufacturing, while the remainder are still in pilot or R&D stages. Although an early stage technology, in large scale solar silicon production the signs look encouraging for increasing quantities of silicon available for the solar sector in future years.

However, upgraded metallurgical silicon could result in reduced cell efficiency, offsetting at least in part some of the cost advantages due to the cheaper raw material. There have been conflicting reports on the exact effect that use of upgraded metallurgical silicon has on cell efficiency, and it has generated a lot of controversy. For example, according to Asia Silicon, even the best upgraded metallurgical silicon (with five nines purity) reduces absolute solar cell efficiency by 1 percent, with a consequent 7 percent margin erosion for PV manufacturers. However, recent developments point to upgraded metallurgical silicon indeed emerging as a possible alternative to industry-standard polysilicon.

For instance, Q-Cells AG (ticker: QCE GR), a German solar cell company, has tested upgraded metallurgical silicon from 16 suppliers and found that as many as 5 of them have been able to produce upgraded metallurgical silicon meeting their quality requirements. Q-Cells has been using upgraded metallurgical silicon in its optimized process at between 50 to 100 percent fill rates. Indeed, in a meeting with Alternative Energy Investing™, Q-Cells management stated that using upgraded metallurgical silicon had little adverse impact on yield and cell efficiency. In its field tests of PV modules using 100 percent upgraded metallurgical silicon, Q-Cells has not found any material power loss.

To give another example to illustrate the potential of upgraded metallurgical silicon, Canada-based Timminco expects to have approximately 15,000 metric tons (mt) of upgraded metallurgical silicon capacity sometime in 2009. Its production process has been certified in an independent assessment by Photon Consulting. During interviews conducted with customers as part of this assessment, Photon found several cases where the cell efficiency was 14 percent plus, and in some cases 15 percent plus, using unblended upgraded metallurgical silicon. Further, Timminco has a robust order backlog, which adds further credibility to its claims on upgraded metallurgical silicon technology. Unfortunately for Timminco and its customers, scaling up of the manufacturing process has met with delays.

To give an idea of the economics of this, upgraded metallurgical silicon priced at $30/kg, and virgin poly priced at $65/kg, using long-term contract pricing for each, results in average wafer costs of $0.83 and $0.89 per watt, respectively, with comparable efficiencies and processing parameters. Therefore, as long as polysilicon prices remain above approximately $55/kg to $60/kg, upgraded metallurgical silicon offers polysilicon-based solar wafer and cell companies an alternative to the higher prices of virgin polysilicon, expanding supplies and potentially keeping polysilicon prices low.

In short, the addition of upgraded metallurgical silicon will help slow price increases by adding significant capacity into the industry, but the economics of upgraded metallurgical silicon would become unfavorable should virgin poly prices fall dramatically below $60 per kg.

Ingot/Block/Ribbon Manufacturing

The polysilicon made using either the Siemens process or other processes needs to be converted into an intermediate product. Depending on the specific process used, the polysilicon is then made into monocrystalline ingots, or polycrystalline blocks, ribbons, or tubes. Let's consider the processes involved for each of these.

Monocrystalline Ingots

The established method to manufacture monocrystalline silicon ingots is the "Czochralski process" (Cz). In this, high-purity polysilicon, with suitable "dopants" (materials such as arsenic or boron, added in order to impart certain electrical properties), is melted in a quartz crucible. A "seed" crystal of silicon is lowered into the melt and then "pulled" at a specific rate away from the melt.

This process of "crystal pulling" is complex and utilizes precise control systems to ensure that crystal growth progresses as required. The process results in the formation of a cylindrical ingot of monocrystalline silicon around the seed crystal. The Cz process, though intricate, is well-established in the industry.

An alternative to the Cz process is the "Float-zone" (Fz) process, which also produces monocrystalline silicon. The Fz process produces ingots that are purer (and hence more efficient) than ingots produced using the Cz process. Monocrystalline ingots are most suitable for the semiconductor industry, which seeks the highest purity and can easily afford to pay higher prices in comparison to the solar industry.

Polycrystalline Blocks

In directional casting, polysilicon is melted in crucibles and cast into polycrystalline blocks. The casting process ensures that the silicon that solidifies is always below the molten silicon, so solidification always proceeds unidirectionally. This is to avoid any structural defects in the blocks, since silicon expands on solidification.

Directional casting is faster and cheaper than the Cz/Fz process of manufacturing monocrystalline ingots. However, solar cells made from polycrystalline blocks have lower conversion efficiencies compared to cells made from monocrystalline ingots.

The process of converting these ingots and blocks into wafers results in a substantial waste of silicon—up to 50 percent. This waste is known as "kerf loss." There are other technologies that bypass the traditional poly-ingot-wafer route and which consequently result in reduced use of silicon. Two such technologies, which produce "ribbons" or "tubes" instead of ingots, are described next.

Polycrystalline Ribbons/Tubes

String Ribbon. String ribbon is a technology proprietary to Evergreen Solar of Marlborough, Massachusetts (ticker: ESLR). It does not require the process of making ingots in the traditional sense. Instead, it involves manufacturing "ribbons" of polysilicon, which can then be laser cut into wafers.

The process starts with melting polysilicon in a graphite crucible. String ribbon technology utilizes a set of parallel wires or strings to "pull" a ribbon of silicon from molten polysilicon, so the melt solidifies between the strings.

Evergreen's Gemini II furnaces are capable of harvesting two ribbons at a time from molten polysilicon. Further, Evergreen has developed state-of-the-art "Quad" furnaces capable of pulling four ribbons simultaneously.

String ribbon has the advantage of requiring the least amount of polysilicon per watt (as of 1Q 2008, less than 5g Si/W, as compared to the industry average of less than ~10g/W).

Edge-Defined Film Fed Growth (Tube Ribbon). A technology patented by Schott AG, in this method an octagonal tube is "pulled" from molten polysilicon so a hollow octagonal silicon "tube" with a thin wall is created. Wafers are produced from this octagonal tube by laser cutting, with very little silicon wasted compared to the traditional ingot/block-to-wafer approach. The process of making wafers with this technology is also more energy efficient.

Wafer Manufacturing

Starting from Ingots

A simplified version of wafer manufacturing at MEMC, an established wafer manufacturer, provides a useful illustration of the basic process. (LDK and Renasola are also in this business.)

The starting material for MEMC's wafer slicing process is a monocrystalline ingot. The initial stage of processing involves removing the ends of the ingot using a saw.

The ingot is then cut into smaller pieces to facilitate slicing. MEMC tests a sample portion of the ingot for key parameters (impurities, electrical characteristics such as resistivity, and so on) and also performs X-ray tests to verify that the crystal orientation is correct. After some preparatory steps, the ingot is mounted on either an ID saw or wire saw. (Meyer Burger and HCT Shaping, a subsidiary of Applied Materials, are in the business of making saws.)

An ID saw can produce only a single wafer at a time. On the other hand, a wire saw is a cutting tool that employs a web of fast-moving wire to slice the ingot all at once. The advantages of a wire saw over an ID saw are:

- It has a lower cycle time because the cutting operation happens all at once.
- "Kerf loss," that is, loss of silicon material during slicing, is lower.

The wafers obtained from the ID or wire saw are then treated chemically to free them of unwanted slurry. These are then "profiled" for the purpose of strengthening. Subsequently, they are marked with lasers so they can be traced to a specific manufacturing batch and then worked on by a "lapping" machine to improve surface features.

Next, the wafers undergo "etching," a process in which they are chemically treated to further improve strength and create a glossy surface. Finally, the wafers are polished and cleaned.

Starting from Ribbons/Tubes

Evergreen Solar uses ribbons obtained from its string ribbon process as the starting material for wafering. It has developed a laser-based automated "cut on the fly" technique that can potentially improve wafer yields. Similarly, Schott AG, whose wafering process starts with octagonal silicon tubes, employs a laser cutting technique that minimizes waste of silicon.

CELL MANUFACTURING

A review of the process of manufacturing cells from wafers using the example of Q-Cells AG illustrates the key step where a silicon wafer gets converted into a solar cell capable of generating electricity.

We first start with a high-level description of the manufacturing process at Q-Cells and then move to a detailed profile of one of the manufacturing lines (Line 4) based on factory visits conducted by Alternative Energy Investing™. A study of Line 4 is particularly useful because it serves as a template for the "finished wafer to cell" configuration at both Q-Cells and the industry.

The basic steps of cell manufacturing at Q-Cells are summarized below:

1. Silicon wafers, arriving from key suppliers and the key input to cell manufacturing, undergo rigorous quality tests (for surface features, dimensions, etc.).
2. Wafers that pass the quality tests are cleaned by being subjected to a wet chemical process, followed by etching and rinsing processes, after which they are dried.
3. In order to create a solar cell, a "p-n junction"— a cell that has a positive p-type semiconductor on one side and a negative n-type semiconductor on the other—needs to be created out of the wafer, which is a pure p-type semiconductor. This is accomplished by diffusing one side of the wafer with phosphorus in a furnace.
4. The intermediate product is tested for its electrical characteristics, and impurities created during the diffusion process are eliminated by another round of etching.
5. A coating of silicon nitride is applied to ensure that as much sunlight as possible gets captured by the solar cell. This is done by exposing the wafer to a high-temperature (~400 degrees Celsius) reaction between ammonia and silane.

6. Now, "contact grids" need to be created on either surface to facilitate the flow of electricity. A "screen printing" technique is employed to "print" thin strips of silver on either side of the cell. Further, the back of the cell (the p-type side) is coated with aluminum.
7. The cells are tested, and quality ratings are assigned to each cell.

Detailed Study of Line 4 Q-Cells
Most of this section is reproduced from a note prepared by Alternative Energy Investing™ to summarize the results of its visits to the Q-Cells' facility ("Applied Materials: Where Does Baccini Play?" November 19, 2007; a subsidiary of Applied Materials, Baccini makes screen-printing equipment). The first 12 steps are as follows:

1. Wafer Unpacking, Visual Inspection, and Loading.   This is a preliminary step where wafers are unpacked from cardboard boxes and tested on key parameters.

2. Etch and Texturing.   The inline conveyor distributes wafers across parallel conveyors entering a Schmid acid texturing chamber. During this process the damage caused by the wire saws during the wafer cutting process is removed, and at the same time the surface is increased. The acid texturing is based on a mixture of HF (hydrofluoric acid) and HNO_3 (nitric acid). With multicrystalline wafers, this results in increased efficiency.

After the actual texturing, the wafers run through a cascaded rinsing process with water followed by a NaOH (sodium hydroxide) rinse, water rinsing process, and then an acidic HF process. There is a final water rinsing process followed by a drying process on the basis of special dry jet modules, leaving the etched and textured wafers streak- and droplet-free. The wafers continue on conveyors into Step 3.

3. Automated Magazine Loading.   As the wafers exit the Schmid drying chamber, they travel on inline conveyors into a Jonas & Redmann automated magazine loader before proceeding on to Step 4 (furnace diffusion) or entering a storage chamber for processing later.

4. Doping and Diffusion.   Step 4 is a gas process where phosphorus atoms are applied to the wafer surface and then diffused with silicon in a batch of $POCl_{3\ in}$ four-chamber Centrotherm furnace to create a SiO_2 layer.

During this step, the wafer is converted from a p-type to an n-type conductor. It is at this point that the cell is created. Line 4 has four "closed tube" diffusion furnaces, which have the advantage of reduced N_2 (nitrogen), O_2 (oxygen), and $POCl_3$ (phosphorus oxychloride) consumption relative to "open tube" furnaces.

5. Phosphorus Test and Measurement.   Phosphorous testing and measurement takes place in Jonas & Redmann resistance equipment. A probe loaded with current touches one of six cells on the line at five points on the surface of the wafer, recording the ohms per square.

6. Phosphorus Silica Glass Etching.   In PSG etching, $CF_4 + O_2$ act aggressively on the cell's edges, creating a clear separation between the phosphorus-endowed negative layer and the cell's positive layer in a Schmid three-chamber wet process. The rinse steps are H_2O.

7. Automated Magazine Loading.   As the cells come out of the PSG etch step, they enter an automatic magazine loader for transporting to Step 8 or into storage for processing later.

8. Antireflection Coating.   To ensure that as little sunlight as possible is lost, the cells are coated with silicon nitride, which imbues a blue color to the cell. Hydrogen in the silicon nitride also contributes to a significant increase in cell efficiency through passivation of defects in the crystal structure.

Before coating, the cells are transferred from the magazines by two Staubli robots into graphite boats. These robots load and unload in parallel uncoated and coated cells. Uncoated cells are subjected to a high frequency field applied to form plasma between the graphite plates. Silicon nitride is deposited in a very fast three-minute step before the boats are then moved into a Centrotherm diffusion furnace operating at between 400 and 450 degrees C. This is a plasma-reinforced vacuum process with ammonia (NH_3) and hydrosilicon (SiH_4). These reaction gases are deposited onto the cell's surface as silicon nitride.

Coated cells are once again loaded into magazines automatically and transported for testing in Step 9.

9. Silicon Nitride Thickness Testing.   An ICOS color camera tests each cell's color, measuring the thickness of the silicon nitride coating. After color testing, Line 4 splits into two parallel lines for screen printing.

10. Screen Printing.   Silver paste is applied first to one side of the cells, using Baccini printers with two printing heads. Once screen-printing of the front side is done, followed by inspection and drying, backside printing is done. This is accomplished in a two-step process.

In the first step, an aluminum and silver layer is screen printed—silver here helps in soldering the contacts. Cells are then tested, and those that pass proceed to receiving a back-side layer of aluminum.

The next step of the screen-printing process is to apply aluminum paste to the back side of the cells screen printed around the soldering lines. This step completes a back-side contact, a reflective metal surface (mirror), and a back-side surface field, concluding the passivation of the back side to prevent some recombination and loss. Upon completing the second and final step in back-side printing, cells are photo inspected before entering a drying chamber. After drying, the cells proceed to the next step, sintering.

11. Sintering.   In order to imprint the cell's contacts and aluminum layer in the wafer, they enter a Centrotherm-firing furnace. There, the etching of the silver into the silicon nitride and the passivation of the front side occurs. Simultaneously, the aluminum diffuses into the back side of the cell. There are no gases involved in this step. Upon completing of the sintering step, cells move through a loading and unloading station where operators can clear broken cells and researchers can load or unload test cells.

12. Test.   Cells then progress into a cell tester, testing for front-side and then back-side (automated cell flipper) geometry. Cells move to the next step to mitigate the damage caused by sintering when phosphorus diffusion occurred and could not be avoided.

When the above 12 steps are finished, there are three more steps. Steps 13, 14, and 15 involve isolation etching, cell testing, and sorting based on key parameters (color, efficiency, and so on).

Module Manufacturing

Module manufacturing involves connecting solar cells electrically by "weaving" conducting material between them to obtain the desired voltage and current characteristics. Cells connected in this manner are enclosed in a package to protect them from the elements. The manufacturing of the module is done in such a way that even if the cells were to move slightly from their positions, the electrical connections remain more or less intact.

A simplified schematic diagram of a solar module (not to scale) is presented in Figure 1.3. The components depicted in the diagram are mounted on a frame made of materials such as aluminum.

The front surface, which is made of materials like glass, is intended to protect the cells. The material used for this needs to be as transparent as possible so the efficiency of the module is maximized.

The encapsulant acts as a sort of glue that holds the module components together. Ethyl vinyl acetate, abbreviated as EVA, is a popular choice for this. The process of manufacturing a module involves inserting EVA sheets as depicted in the schematic and heating it, which causes EVA to bind the components together.

The back surface is made of materials like Dupont's Tedlar (polyvinyl fluoride), and serves to protect the enclosed solar cells.

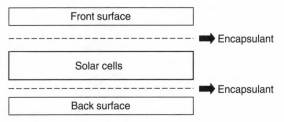

Figure 1.3 Diagram of a solar module

The degree of automation for module manufacturing is specific to each plant. There are plants (such as SunPower's Philippines-based panel manufacturing plant) that have a high degree of automation, and other plants that are more labor intensive and do not have such highly automated processes, like many of the Chinese module companies (Suntech and Canadian Solar, which is incorporated in Canada but conducts all of its manufacturing operations in China). We will return to automation later in the book, when we address the question of comparative advantage.

USING PRODUCTION METHODS
TO PICK STOCKS

Now that you have a working knowledge of just how polysilicon solar is made from beginning to end, the question arises: How can you use this knowledge to pick stocks? As illustrated in the pie chart in Figure 1.2 on page 12, polycrystalline is the largest technology segment, with monocrystalline a strong second, making the overall crystalline silicon group the dominant force in today's market. But this dominance is being challenged by a growing number of companies testing, commercializing, and marketing their new technology. Knowing how to judge which companies among the incumbents will outmaneuver the others and understanding how to diversify technology and risk are keys to investing in solar stocks.

Within the crystalline silicon group we currently find three general groups of companies vying for supremacy. The first category designs for more efficient cells (for example, Sunpower), with the aim to demand a price premium. The second group puts a premium on cost reduction strategies with wafer technology that uses less silicon (Evergreen Solar). And last, there are companies without much differentiation that comprise a group within crystalline silicon category that we call "followers."

These three groups create a simple framework for investors to evaluate the investment prospects of any crystalline silicon company. It is critical to know which company is about to get bypassed by a challenger using a new strategy or by one that does not use silicon at all (like thin film companies such as First Solar). Ultimately, the successful investor will diversify risks coming from inside or outside the crystalline silicon group of solar companies.

Thin Film Solar Technology

We believe that in the near future the crystalline silicon group of companies will lose share to incumbent and emerging thin film technologies. Thin film currently accounts for 9.99 percent of the market, but this share is expected to grow to 20 percent by 2010 (Alternative Energy Investing™). For reasons that will become clear as we discuss the technology, thin film holds significant advantages in manufacturing processes, which in turn leads to the all-important lower costs per kilowatt hour.

Rather than growing, slicing, and treating a crystalline ingot as with crystalline silicon, photovoltaic material can be created by sequentially depositing thin layers of different materials into a very thin structure. The resulting thin film devices require very little silicon material, if any, and have the added advantage of being capable of economies of scale. Several different deposition techniques are available, all of them potentially cheaper than the techniques required for crystalline silicon. Best of all, these deposition processes may be scaled up so the same technique used to make a small laboratory cell can be used to make a very large module. Thin film companies, both public and private, are essential to successfully investing in solar stocks. Understanding who, what, when, and where among them is indispensable.

Manufacturing Process for Thin Film Solar

The manufacturing process for thin-film-based solar in essence involves depositing thin layers (usually 3 to 5 microns versus a crystalline silicon wafer with a thickness of ~180 microns or more.) of semiconducting material on a substrate. The manufacturing process for thin film (TF) has the following advantages, compared to the process for crystalline silicon:

- Commercial deposition techniques can be employed to deposit the photovoltaic semiconductor material on a substrate.
- The process flow from deposition to assembly and testing of TF modules can be highly automated. Further, continuous process technology can be utilized for thin film manufacturing, which means higher throughput and potentially lower module manufacturing costs.

Thin film photovoltaics could be of several types, the most common of which are based on cadmium telluride (CdTe), amorphous silicon (a-Si), a-Si (μm), dual junction microcrystalline silicon/amorphous silicon ("micromorph") , and copper indium (gallium) selenide (CIS/CIGS).

Table 1.2 gives examples of solar companies that employ each type of technology.

T A B L E   1.2

Technologies Employed by Various Solar Companies

TF Technology	Companies Employing the Technology
Amorphous Silicon: a-Si	AOS Solar, Arise Technologies, Bangkok Solar, CG Solar, CSG Solar, EPV, Equation (Solar Morph), ErSol Thin Film GmbH, Formosun Technology, Free Energy Europe, Grupo Unisolar, Heliodomi S.A., HelioGrid, ICP Solar Technologies Inc, Intersolar UK, Kaneka Corporation, Kenmos, Lambda Energia, Malibu, Mitsubishi Heavy Industries, Moser Baer, MWOE Solar, NexPower Technology (UMC Group), QS Solar (Nantong Qiangsheng Photovoltaic Technology Co), Schott Solar, Signet Solar, Sinonar Corporation, Solar Cells, SolarPlus, Sunner Solar, Sunwell Solar (CMC Magnetics corp.), T-Solar, TerraSolar Inc, Tianjin Jinneng Solar Cell Co.,Ltd, Titan Energy Systems Ltd., Topray Solar, Trony Science, United Solar Ovonic (UNISolar), VHF-Technologies, XsunX, Xunlight
Amorphous Silicon (Tandem): a-Si (μm)	Auria Solar, Bharat Heavy Electricals Limited (BHEL), Fuji Electric Advanced Technology, Inventux Technology AG, Kaneka Corporation, Pramac SpA, Sharp, Sontor GmbH, SunFilm Solar, Suntech Power Co., Ltd, Sunways AG, VHF-Technologies
Copper Indium Gallium Selenide: CIGS	NanoSolar, Ascent Solar Avancis, CIS-Solartechnik, Daystar (DSTI), Global Solar Energy Inc., HelioVolt, Honda, Johanna Solar, Miasolé, Nanowin Technology, Odersun AG, PV Flex Solar, Scheuten Solar, Showa Shell, Solarion, Solibro AB (Q-Cells), SoloPower, Solyndra, Sulfurcell Solartechnik, Titan Energy Systems Ltd., Würth Solar
Cadmium Telluride: CdTe	First Solar, Antec Solar, AVA Solar, Calyxo (Q-Cells), Sunovia, PrimeStar Solar

Dr. Michael Powalla of ZSW (a German research foundation) notes certain manufacturing process characteristics for each type of thin film technology. In particular, he observes that a-Si technology can be implemented through turnkey solutions, while CdTe has the advantage of fast processing and CIS is relatively complex.

The following examples of three companies illustrate the manufacturing processes for thin film solar.

Uni-Solar

In general, a-Si can be deposited on two kinds of substrates: glass and flexible materials such as stainless steel. A subsidiary of Energy Conversion Devices (ticker: ENER), Uni-Solar, manufactures a-Si thin film laminates on a stainless steel substrate, and has managed a significant degree of success in the building integrated photovoltaics (BIPV) market.

Uni-Solar uses a low temperature vapor deposition process in a vacuum that requires much less energy than is required for crystalline-silicon-based photovoltaics. Uni-Solar also has a patented "continuous roll-to-roll solar cell deposition process." Roll-to-roll processing ensures that costs related to process setup are minimal, and it also enhances uniformity of deposition. The company's manufacturing facility in Auburn Hills, Michigan, employs a high degree of automation to produce thin film photovoltaics around one and a half miles long.

Uni-Solar uses a "multijunction" approach to ensure that as much of the sun's energy as possible gets captured. In particular, this approach involves depositing multiple thin layers of semiconductor material, so there is one cell each for capturing the blue, green, and red portions of the light spectrum. To be specific, a-Si alloy material is used to capture blue light, a-SiGe alloy for green, and a-SiGe alloy for red light.

Further, Uni-Solar's module design uses a "bypass diode" corresponding to each cell, so there is minimal loss of power output in conditions of partial shade. Uni-Solar's flexible laminates currently use EVA and DuPont's Tefzel as encapsulants and are easily integrated on commercial roof tops.

First Solar

One of the noteworthy success stories of thin film manufacturing is First Solar, which has been able to reduce manufacturing costs to just $0.99 per watt as of 4Q 2008. First Solar employs a highly automated manufacturing process flow from deposition to assembly and testing of thin film modules. The manufacturing of First Solar's CdTe-based cells takes place on a continuous line with the following stages:

1. *Deposition.* In this stage the glass substrate is first cleaned and heated. Then a proprietary "vapor transport deposition" technology is used to deposit CdS (cadmium sulfide), followed by CdTe (cadmium telluride). The deposition is followed by a rapid cooling process in order to increase strength.

2. *Cell definition.* This stage employs a proprietary "laser scribing" technology to convert the CdS/CdTe-coated glass into photovoltaic cells.

3. *Assembly and test.* This stage involves making the necessary electrical connections to make a module and enclosing it using suitable laminates, frames, and so on. This is followed by module testing. This is the only stage in First Solar's manufacturing process that is not completely automated.

NanoSolar

NanoSolar uses a "printing" process to deposit copper indium gallium selenide (CIGS) ink on a substrate of conducting metal foil. This ink uses a proprietary nanotechnology-based process that ensures the semiconductor ink is homogeneous. NanoSolar's process technology also allows for continuous manufacturing and has the added advantage of not requiring a vacuum deposition chamber.

The common thread we observe across the three TF companies whose manufacturing processes we have studied is that unlike the batch-based manufacturing processes for crystalline silicon, thin film approaches can highly automated and continuous. As Tom Feist of GE Research's Thin Films Laboratory observes, "The challenges that thin film manufacturing faces at the moment are related to arriving at roll-to-roll and lower temperature processing, and obtaining higher throughput." As TF technology is further commercialized in coming years with potentially lower cost manufacturing methods compared to today, investors must select those companies that can prove they can scale their process, decreasing their costs while managing to consistently improve efficiencies along the way.

Micromorph Thin Film

Another emerging technology that could prove significant in the future is "micromorph," which involves depositing a layer of amorphous silicon and a layer of microcrystalline silicon on glass. The amorphous silicon layer is meant to absorb the visible part of the electromagnetic spectrum, while the microcrystalline layer is meant to absorb the infrared and near infrared portions. According to Q-Cells, this helps increase module efficiency to 10-percent-plus levels, compared to 6–8 percent for stand-alone amorphous silicon modules. Sontor GmbH, a subsidiary of Q-Cells, manufactures micromorph-based TF in partnership with Applied Materials on the equipment side, and Julich Research on the R&D side.

Oerlikon, a Swiss-based equipment manufacturer, describes micromorph as having a "tandem structure with an additional microcrystalline absorber."[1] Oerlikon also states that conversion efficiency can be boosted by up to 30 percent using this technology.

Potential Drawbacks of Thin Film Technology

While thin film holds great promise, investors should take into consideration that the technology holds several risks that may limit its commercial viability. As with everything in the investment world, not everything is as it seems. Thin film technologies are leading the way to grid parity, but their lead is not assured, nor is it immune to changes in the cost structure for polysilicon-based PV brought on by vertical integration, improvements throughout the polysilicon-based PV supply chain bringing about substantially lower costs, or improvements in technology from development projects around the world.

The greatest of these risks is thin film's relatively high *installed system cost per watt*. While thin film holds seemingly insurmountable advantages at the module level, conversion efficiency differences and technology maturity paint a very different picture at the system level. Installed TF systems must be larger to compensate for thin film's lower efficiency, which translates into higher costs.

The enormous advantage CdTe has at the module level shrinks to roughly 10 to 15 percent at the system level.

Technology differentiated thin film module manufacturers should drive module cost per watt lower as a result of volume production and continuous manufacturing improvement, begin to integrate downstream to the end customer to sell energy under a distributed model when regional economics dictate, and, depending on the present technology used, evaluate new or additional technologies to supplement present technologies in down years. (This would make sense for a company that could face material shortages—like Te, In—within the next 5 to 10 years.)

CdTe and high-efficiency c-Si are early leaders in the commercialization of solar PV, but these technologies will face competition from a-Si (um) and CIGS/CIS companies that have yet to prove they can commercialize their technologies. Indeed, CIGS could challenge industry leaders in several years, but has yet to see volume production. Many of these companies will attempt to commercialize 20 MW to 40 MW lines in 2009. As CdTe, tandem cell, and CIGS/CIS claim the ascendancy in the thin film segment, the oversupply developing in the market due to the onslaught of new polysilicon supplies should be far less likely for these companies. Because thin film companies are a far smaller percentage of the industry, TF modules should not experience the same kind of periodic oversupply that plagues polysilicon-based modules. Established TF module manufacturers—and there are not that many—will enjoy a more insulated ride through an industry shakeout over the next few years as their low costs and improving efficiencies potentially protect their lead or extend it. It is also useful to add that in cases such as UniSolar's flexible laminates, despite lower conversion efficiency compared to c-Si technology, their lack of mounting structures may put TF frameless products on par with c-Si in certain rooftop systems.

RESEARCH AND DEVELOPMENT OUTLOOK

Understanding the battle between crystalline silicon and thin film gives investors a good idea of where the market stands at the moment, and where it will likely be heading in the near future. But to understand the long term, investors must look at research and development. Although R&D is a key factor in maintaining technological advantage over competitors, the following section is not intended to whet the investor appetite but rather to make note of very smart minds with a notion to upset the status quos sometime in the future. Here are some current R&D efforts of which you should be aware.

Europe

A team of scientists from all over the United Kingdom, led by the experts at Durham University, are undertaking one of the largest photovoltaic solar energy research projects. The project, PV-21, is estimated to cost around £6.3 million and will focus on making thin film light-absorbing cells for solar panels from sustainable and affordable materials. The project started in April 2008 and is expected to finish by 2012. It is being funded by the Engineering and Physical Sciences Research Council (EPSRC) under the SUPERGEN initiative. It will work in coordination with nine industrial partners toward a goal of making solar energy more competitive and sustainable (http://www.azonano.com/News.asp?NewsID=5654).

United States

The U.S. Department of Energy (DOE) recently announced $22.7 million in grants for the basic research projects in the solar PV sector aimed at improving the capture, conversion, and use of solar energy. This research will help increase the contribution of solar power in U.S. total energy supply. The project is being carried out on a large scale:

- DOE's Office of Science selected 27 projects that will focus on fundamental science to support enhanced use of solar energy.
- Universities and national laboratories in 18 states will conduct the research.

The projects are part of a departmentwide, comprehensive, balanced portfolio of basic and applied research and technology development aimed at significantly advancing the use of sunlight as a feasible solution to meet the nation's compelling need for clean, abundant sources of energy.

DOE had plans to fund additional projects starting in 2008 that would focus on research for the conversion of solar energy to electricity. Approximately 14 such projects are now in line, with a total investment of $9.9 million over three years. The challenge in converting sunlight to electricity is to greatly reduce the cost per watt of delivered solar electricity by dramatically improving the conversion efficiency.

A broad range of research on novel approaches to solar-to-electricity conversion is covered by these projects, including:

- Nanostructured inorganic photovoltaics
- Plasmonic conversion concepts
- Organic and hybrid inorganic-organic photovoltaics
- Multiple-exciton generation for enhanced conversion
- Nanoarrays for improved photoelectro chemical cell performance

Here's a brief explanation of some of the key innovations[2]:

- *Nanoarchitecture* involves the use of materials at a scale of 1 to 100 nanometers (1 nanometer = 10^{-9} meters) to help generate electricity.
- *Multiple exciton generation* raises the theoretical conversion efficiency of single-junction solar PV cells from 33.7 to 44.4 percent. This technology essentially involves generating multiple excitons for each photon with adequate energy.
- *Plasmonics* involves the use of nanomaterials so more light enters the absorber, thereby improving solar cell performance.

R&D improvements could be related to both technology and process. We take SolarWorld as an example for this and describe some of the company's key R&D initiatives (SolarWorld 2007 annual group report, p. 77).

- Reducing material loss in sawing by up to 20 percent
- Reduction in wafer thickness
- Reduction of raw material used for wafering

LDK Solar has a detailed, metric-based road map of goals it seeks to achieve by 2013 (LDK Solar 2008 analyst day presentation, San Francisco, July 16, 2008):

- Increasing ingot weight from 450 kg to 1000 kg
- Reducing energy consumption for making ingots from 8 kWh/kg to 6 kWh/kg
- Increasing wafer size from $156 \times 156\,\mathrm{mm}^2$ to $210 \times 210\,\mathrm{mm}^2$
- Decreasing wafer thickness from 180 μm to 120 μm
- Reducing kerf loss
- Reducing silicon consumption from six to eight grams per watt to four to five grams per watt
- Increasing efficiency of cells derived from LDK's wafers, from 15.8 to 18.0 percent

LDK Solar also has four different types of "next generation" wafers to leave it well positioned for the future:

- Terra wafers: N-type monocrystalline wafers to achieve high efficiency
- Nova wafers: polycrystalline, based on upgraded metallurgical silicon
- Bright wafers: polycrystalline, with enhanced quality
- Aurora wafers: polycrystalline, with a differentiated casting process

First Solar has two main thrust areas where its R&D efforts are focused (10-K filing for FY07, p. 7):

- *Increasing conversion efficiency.* First Solar focuses on ensuring that the maximum number of photons reaches the absorption layer and that the maximum number of electrons reaches the CdTe surface. It also focuses on mitigating electrical losses.
- *Optimizing the system.* First Solar adopts an approach where it seeks to optimize the entire solar PV system. To this end, it obtains data from installations that use its modules and uses this as a basis for improvement.

Energy Conversion Devices is a company that has a long tradition of investing in R&D and coming up with major technology breakthroughs. Currently, its research efforts for solar photovoltaics focus on:

- Finding ways to reduce production costs
- Improving conversion efficiency
- Innovating on the marketing side, as well as identifying new applications for its products

Evergreen Solar focuses its research and development on several fronts, which include:

- Improving its proprietary string ribbon technology so its silicon consumption will be reduced to just 2.5 grams per watt by 2012
- Improving conversion efficiencies
- Improving factory yields

JA Solar has taken the following steps to innovate in the area of research and development:

- Increasing conversion efficiency
- Reducing production costs

As part of this effort, JA Solar has been setting up research facilities in Yangzhou, and also through its wholly owned subsidiary in the United States, JA USA.

Trina Solar has R&D activities focused on the following three major areas[3]:

- Innovating in manufacturing processes across the solar PV supply chain for manufacturing monocrystalline ingots, wafers, and cells, as well as modules
- Obtaining knowledge on how to use more of reclaimable silicon as raw material for making monocrystalline ingots and wafers
- Creating a "platform" for using reclaimable silicon to be used for the production of polycrystalline silicon

China Sunergy focuses its R&D efforts on the following three aspects:

- Increasing conversion efficiency
- Developing new products such as N-type solar cells (these usually give better conversion efficiencies than the P-type cells that are normally used)
- Formulating plans to develop passivated emitter and rear cells in the future

We see that most companies have a focus on improving conversion efficiency as part of their research and development efforts. This is important because for every percentage point of improvement in conversion efficiency, we will be able to decrease installed system costs (residential, United States) by ~5 percent (SunPower, annual report, 2007).

N O T E S

1. http://www.oerlikon.com/ecomaXL/index.php?site=SOLAR_EN_thin_film_si_solar_moduls
2. "Next Generation Photovoltaic Devices and Processes Selections,, U.S. Department of Energy solar energy technologies program, November 8, 2007
3. Trina Solar, 20F filing for FY07, p. 46

CHAPTER 2

Monitoring the Industry

SOLAR: A GLOBAL MARKET

As mentioned in Chapter 1, solar PV generated electricity is more expensive than electricity produced through the conventional means of coal, natural gas, and nuclear power plants. While great progress has been made in recent years to bring solar to grid parity, it will remain more expensive for at least the near future in most major markets. This price barrier would perhaps be an insurmountable challenge if not for the fact that in an effort to stimulate demand for solar PV systems, some governments, at both the national and local levels, have voted in or mandated renewable energy programs employing various financial incentives.

These incentives underpin the solar PV industry and provide support and demand for solar PV electricity that otherwise would scarcely have a market. So it can be seen that understanding how these incentive and subsidy programs work is an important part of a successful solar investment strategy. This chapter will cover the structure of various incentive programs and how governments, utilities, and consumers are putting them into practice, as well as their effects on the global solar market. It will also consider the role that state-mandated Renewable Portfolio Standards (RPS) and the federal investment tax credit (ITC) play in driving the growth of the solar industry.

NET METERING AND FEED-IN TARIFFS

Two of the most widely used programs are feed-in tariffs (FIT) and net metering. Each program aims to offset the cost of solar PV systems by providing incentives for consumers and businesses that install them.

A feed-in tariff sets a fixed, above-market rate that utilities pay to purchase the solar electricity generated by a consumer or company. For example, if the retail rate for electricity in a particular market is 10 cents per kilowatt hour, the feed-in tariff for that market might be 40 cents per kilowatt hour. In some countries the utilities receive certain incentives for purchasing solar electricity and other forms of renewable energy; such incentives become important considering that the tariff utilities pay to the homeowners for solar electricity is set at a higher rate than the grid produced electricity. The feed-in tariff rate is set high in order to encourage homeowners to participate and also to ensure that they recoup their investment in the costly solar installation.

Germany is the leader in feed-in tariff policy with its Erneuerbare Energien Gesetz (EEG) law, which was first introduced in 1990 and remains a model for many other countries. By 2007, 46 governments across the world had feed-in tariff laws, including the government of Spain, at the national level, and the Australian territorial governments of Queensland and South Australia. In February 2008 the state of California also approved a feed-in tariff applicable for up to 480 MW of solar energy created by small producers.[1]

As for net metering, these are programs that reward consumers and companies that are generating solar electricity by deducting their system's solar electricity outflow from the grid electricity inflow into that house or facility to find the net balance; hence the name "net" metering. The solar PV system owner receives a retail credit from the utility for that extra portion of electricity generated after the deduction. Usually this is calculated through the system owner's electricity meter spinning backward for the solar electricity generated. Net metering rules are implemented at the state level and vary widely from state to state, just like feed-in tariffs vary from place to place.

Net metering pricing schemes are especially effective in targeting small energy producers that produce more solar electricity than they consume. According to the EIA, over 34,000 customers in the United States utilized net metering programs in 2006, with the vast majority (about 75 percent) of these customers located in California. The number of customers accessing net metering programs represents about 1 percent of all the customers in the United States.

Feed-in Tariffs vs. Net Metering

In contrast to net metering, feed-in tariffs allow producers of renewable energy a certain return on investment without the volatility of wholesale and retail market pricing. Under the net metering system, it is difficult to recover their net metering credits and/or rebates (depending on that locale's net metering rules) for using renewable energy.

The rates paid under feed-in tariffs are set higher than standard electricity prices; they are effectively subsidized by the utilities that purchase the electricity, and indirectly by the utility's consumers. Net metering, on the other hand, treats both the solar electricity and the grid electricity at par by applying the deduction rule. This also makes the return calculations very difficult for those individuals who consider installing the solar PV system, in comparison to the feed-in tariff, where the return on investment calculation is very easy.

A SURVEY OF INCENTIVE PROGRAMS AROUND THE WORLD

Given the importance of incentive programs, it's not surprising that the major markets for solar companies are countries with a robust, attractive incentive system in place for promoting solar PV. Indeed, Germany and Spain, which have favorable feed-in tariffs for solar, are the biggest markets for solar PV. In 2007, out of 2.4GW$_p$ of total global solar installations, ~1.1GW$_p$ (46 percent) was in Germany, and ~425 MW$_p$ (18 percent) was in Spain (presentation on June 19, 2008 by Thomas Chrometzka, international affairs officer, BSW-Solar).

The following country-by-country breakdown of incentive programs will be a useful primer for investors as they begin to evaluate solar stocks. Keep in mind, however, that many of the following programs could change as local politics dictate.

Germany's FIT

In June 2008, Germany adopted a revised version of the Renewable Energy Sources Act (EEG) to supersede the earlier version, which was in existence since August 2004. This act has been the key driver behind Germany's status as a leader in the push toward renewable energy; new PV installations in 2007 amounted to around one-half of the total world's installations for the same period.

The EEG law provides for feed-in tariffs for electricity supplied to the grid, with access to the grid guaranteed by law if the necessary conditions are satisfied. Different feed-in tariffs have been fixed for different sources of renewable energy. In addition, some German states also have other subsidy programs to supplement the feed-in tariff program.

For solar, the feed-in tariff is constant for 20 years from the date of commissioning. To illustrate, for roof-mounted systems commissioned in 2007, the feed-in tariff was fixed at ~49 euro cents/KWh, and in 2008 the feed-in tariff was fixed at ~46.75 euro cents/KWh.

To put the feed-in tariffs into a context that investors will find useful, we conducted a sensitivity analysis to calculate the internal rate of return (IRR) for a 20-year period for different installed costs/W_p for a system in Frankfurt. The IRR measures the series of cash flows through the 20-year period that a solar PV system would generate.

To calculate the IRR, we first need to know how much electricity is generated by a solar system each year. This in turn depends on the insolation (the measure of solar radiation energy received on a given surface at a given time) of the specific place we are considering, and varies across different months. In common sense terms, the more sunlight there is in a particular market, the more energy installed solar systems in that market can generate. For this analysis, we used Frankfurt as our hypothetical market. We obtained data about insolation using a Web-based energy production calculator provided by the U.S. National Renewable Energy Laboratory. This tool, called PVWatts, performs its computations for crystalline silicon-based solar cells using certain standard assumptions.

We can observe in Table 2.1 that for the installed cost/W_p of €4.00, the IRR is 7 percent, and as one would expect, as the installed cost/W_p increases, the IRR decreases. The IRR decreases to 1 percent for the installed cost/W_p of €6.50. The hurdle rate—the

T A B L E 2.1

IRR Calculation for One KW Partially Integrated Roof-Mounted System

	Installed Cost/Wp	€4.00	€4.50	€5.00	€5.50	€6.00	€6.50
	0	−€4,000	−€4,500	−€5,000	−€5,500	−€6,000	−€6,500
Year	Revenue per year (20 years)	€375	€375	€375	€375	€375	€375
	IRR	7%	5%	4%	3%	2%	1%

minimum IRR required by German project investors—was approximately 7 percent through the beginning of 2008. The hurdle rate, however, has risen together with interest rates and scarce capital resources used to fund new projects.

EEG specifies a "degression schedule" for feed-in tariffs, which means that the 20-year, fixed feed-in tariff decreases, or degresses, as the date of commissioning is pushed further into the future. Simply put, the later the system is installed, the less the owner receives in tariffs. This feature of the EEG accounts for cost reduction due to learning effects. It also encourages reduction in installed systems costs, so costs per kilowatt hour keep approaching grid parity, at which point incentives can be done away with.

According to the August 2004 version of the EEG, the degression rate for building integrated solar (BIPV) was at 5 percent, and it was at 6.5 percent for ground-mounted systems. But the incentive system in Germany changed greatly over the last few months of 2008. A progress report prepared on the EEG by the Federal Ministry for the Environment, Nature Conservation, and Nuclear Safety came up with recommendations to amend the law. Submitted in 2007, the rationale for the following revisions was that PV system prices (as fixed by the current form of the EEG law) did not adequately reflect the costs of PV systems.

1. It was proposed to increase the degression rate on roof-mounted systems from the current 5 percent to 7 percent in 2009 and 2010, and a further 1 percentage point to 8 percent in 2011 and later.

2. For non-roof-mounted PV, it was proposed that the degression rate be increased from the current 6.5 to 7 percent from 2009, and 8 percent from 2011. This would be backed by a (one-off) one euro cent/KWh cut in tariffs from January 1, 2009, applicable to all PV systems.
3. It was also proposed to create a separate category for roof-mounted PV of capacity >1 MW_p, which would have a feed-in tariff of ~34.5 euro cents/KWh.

Based on these recommendations, the German Bundestag (parliament) adopted an amended version of the EEG in June 2008, applicable for installations from January 1, 2009.

As expected, degression rates were raised. Specifically, for rooftop PV systems up to 100 KW_p, the digression rate is now 8 percent in 2009 and 2010, and 9 percent in 2011. For rooftop PV systems above 100 KWP, the digression is 10 percent in 2009 and 2010, and 9 percent from 2011.

For ground-mounted PV, the degression schedule for 2009 and beyond is similar to that for rooftop-mounted systems above 100 KW_p. In other words, the digression is 10 percent each for 2009 and 2010, and 9 percent for 2011. Further, the new version of the EEG does not have any "bonus" incentives for facade-based PV.

A new feature of the 2008 version of the EEG is the provision of a "sliding scale" for degression. This means if the annual PV megawatt installation in a given year is higher than a stipulated amount, the digression rate would be raised by a percentage point. However, if the annual installation is less than a lower threshold, the digression rate would be lowered by a percentage point. The lower and upper thresholds for this are 1,000 and 1,500 MW_p, respectively, in 2009, 1,100 and 1,700 MW_p in 2010, and 1,200 and 1,900 MW_p in 2011. This gives the German incentive system some degree of flexibility in controlling the growth of PV installations.

For reference, the following tariffs apply for roof-based photovoltaic plants installed in 2008:

- Up to 30 KW_p: 46.75 euro cents/KWh
- 30 to 100 KW_p: 44.48 cents/KWh
- >100 KW_p: 43.99 cents/KWh

Spain

Spain is another major PV market driven by an attractive subsidy program. The original feed-in tariff system was adopted in 2004, and it was revised in May 2007 and is referred to by the identifier 661/2007. The Spanish subsidy program is also referred to as the "Real Decreto," or "royal decree."

In its current form, the Spanish subsidy decree guarantees feed-in tariffs at a fixed rate per kilowatt hour for systems commissioned in a particular year, for a 25-year time period. Further, unlike Germany, Spain does not have an explicit degression schedule, nor can changes in tariffs occur with *retrospective* effect.

To illustrate, for roof-mounted systems commissioned in 2008, the feed-in tariff was fixed at ~44.04 euro cents/KWh.

Similar to Germany (Frankfurt), we conducted a sensitivity analysis and calculated the IRR for the 25-year period for different installed cost/W_p for a system in Madrid. We can see in Table 2.2 that for the installed cost/W_p of €4.00 in Spain, the IRR is 14 percent, which is double that of the 7 percent IRR in the case of Germany. Even if the installed cost/W_p increases to €6.50, the IRR decreases to only 7 percent. The high IRR in Spain can be attributed to the higher insolation in Spain (1,285 KWh of electricity per year for a one KW installation, as compared to just 802 KWh in Germany).

Under the Spanish subsidy program, access to the electricity grid is guaranteed if the necessary conditions are met, and tariffs

T A B L E 2.2

IRR Calculation for One KW Partially Integrated Roof-Mounted System

	Installed Cost/Wp	€4.00	€4.50	€5.00	€5.50	€6.00	€6.50
	0	−€4,000	−€4,500	−€5,000	−€5,500	−€6,000	−€6,500
Year	Revenue per year (25 years)	€566	€566	€566	€566	€566	€566
	IRR	14%	12%	10%	9%	8%	7%

for solar are fixed according to the size of the installation. The following tariffs pertain to the year 2007:

1. For installations ≤ 100 KW, ~ 44 euro cents /KWh (5.75 times the average electricity price) for the first 25 years, ~35 euro cents thereafter
2. For installations > 100 KW but ≤ 10 MW, ~42 euro cents/ KWh for the first 25 years, then ~33 euro cents/KWh
3. For installations >10 MW but ≤ 50 MW, ~23 euro cents/ KWh for the first 25 years, then ~18 euro cents/KWh

The policy is usually renewed every four years, so given that the last revision was done in 2007, we would normally expect the existing policy framework to remain in place until 2011. However, in 2008, due to grossly generous subsidies, Spain reduced its subsidy amounts and placed a relatively low cap on annual installations. During 2008 it is estimated that upward of 2.4 GW of installations were completed in Spain illustrating how poorly conceived subsidy programs that overstimulate can lead to perverse outcomes and unintended consequences. As was the case in Spain in 2008, solar stock investors must recognize when a market is being overstimulated to judge if the poorly conceived subsidy program will swing supply and demand of modules and cause a temporary uptick in profits.

Despite the generous subsidy regime in Spain in the past several years, administration-related issues led to a delayed market response. As Ernesto Macias of Isofoton (a Spanish PV company) noted in a presentation in 2007, there seem to be issues associated with bureaucratic procedures involved in setting up a PV system, getting it grid-connected, and so on. These procedures take a lot of time and require permits from various government agencies and councils, and they act as a dampener to the growth of small scale PV installations. Later that year the government acted to streamline the process leading to a run on modules and an unprecedented boom for solar that briefly benefitted the solar stock investor.

Italy

Italy's law for promoting solar PV is based on a feed-in tariff regime and was revised by a decree in February 2007. As with

Germany, the tariffs are fixed for a 20-year time period. The tariffs for fully building-integrated PV (BIPV) installations in 2008 are: 49 euro cents/KWh for 1 to 3 KW_p, 46 euro cents/KWh for 3 to 20 KW_p, and 44 euro cents/KWh for >20 KW_p. For partially integrated installations, the numbers are 44, 42, and 40 euro cents/KWh, respectively. Finally, for installations that are "not integrated," the tariffs are 40, 38, and 36 euro cents/KWh, respectively. We can see here that the Italian law most encourages BIPV installations in terms of tariffs.

The degression of tariffs is 2 percent for 2009 and 2010. Another aspect to note about the revised law is that it simplifies administrative issues relating to PV, enhancing the potential for Italy to become a major player in the PV market. Further, the program's cap on solar PV installations now stands at the increased value of 1.2 GW_p.

As mentioned earlier, for roof-mounted systems commissioned in 2008, the feed-in tariff was fixed at ~44 euro cents/KWh. Table 2.3 shows IRR calculations for Italy.

Similar to Germany (Frankfurt) and Spain (Madrid), we conducted a sensitivity analysis for Italy (Naples) and calculated the IRR for the 20-year period for different installed cost/W_p. As can be seen in the table, the IRR is 11 percent for the installed cost/W_p of €4.00. That lies in between the values of IRR for Germany and Spain. This is to be expected because the insolation of Italy is somewhere between Germany and Spain. As the installed cost/W_p increases, the IRR decreases considerably to 4 percent for the installed cost/W_p of €6.50.

T A B L E 2.3

IRR Calculation for One KW Partially Integrated Roof-Mounted System

	Installed Cost/Wp	€4.00	€4.50	€5.00	€5.50	€6.00	€6.50
	0	−€4,000	−€4,500	−€5,000	−€5,500	−€6,000	−€6,500
Year	Revenue per year (20 years)	€499	€499	€499	€499	€499	€499
	IRR	11%	9%	8%	7%	5%	4%

United States

The United States has a mix of federal and state incentives for solar.

Nationally, commercial installations are eligible for the Federal Modified Accelerated Cost-Recovery System (MACRS), which allows depreciation over a five-year period. Further, there is a 30 percent federal tax credit (ITC) applied against system cost for residential installations. In 2009, what was commonly referred to as the "stimulus package" lifted the cap on the ITC amount a person or business can take and removed an exemption prohibiting utility companies from claiming the ITC. Other short-term changes to solar subsidy law in the U.S. stimulus package should create a healthy and growing market in the United States that may help to balance the market for module supply and demand in the coming years.

As for the states, they have widely different incentive schemes for promoting renewable energy in general and solar in particular. According to DSIRE (the Database of State Incentives for Renewables and Efficiency), 25 states have rebate programs and performance based incentives (PBIs), and several other states have different mechanisms to promote photovoltaics. Some states have Renewable Portfolio Standards as a basis for promotion of renewables, including photovoltaics. Examples are Maryland, which has a target of 2 percent (of total electricity generation) solar by 2022; Delaware, 2 percent by 2019; Colorado, 0.8 percent by 2020; and California, 20 percent by 2010 and 33 percent by 2020. Also, 16 states have tax credits ranging from 10 to 50 percent.

California is the leading state as far as solar is concerned. There, solar PV has received a major boost since the introduction of the signature Million Solar Roofs program by Governor Arnold Schwarzenegger. The $3.3 billion program aims to create 3,000 MW of incremental solar electricity by 2017.

The $2.2 billion California Solar Initiative (CSI) is a large part of the attempt to achieve this. Administered by the California Public Utilities Commission (CPUC), it lays down incentive schemes to promote widespread adoption of PV.

- For PV systems greater than or equal to a certain size (fixed at 100 KW in 2007, 50 KW in 2008, and 30 KW starting in 2010), incentives will be paid on a monthly basis according to the actual energy produced, for a period of five years. This is called a "performance-based incentive," or PBI, since the incentive is based on actual system performance. (It is important to note that BIPV installations of all sizes necessarily fall under the PBI category.)

- For systems less than a certain size (100 KW in 2007, 50 KW in 2008, and 30 KW starting in 2010), the incentive is in the form of an upfront, onetime payment based on *expected* system output. For residential and commercial installations, the incentive was up to $2.50 per watt in 2007 (the exact number being dependent on expected system performance, which in turn is based on location, orientation, etc.). For installations by government and nonprofits in 2007, the incentive was capped at a higher value of $3.25 per watt. Since this incentive is based on expected performance, it is referred to as an "expected performance based buy down," or EPBB.

Compensation through both PBI and EPBB would be reduced in 10 "steps," based on target installed capacities (for the entire state) being reached. For example, when installations reach the 130 MW mark, the EPBB payment for residential PV will be $1.90/W and the PBI payment $0.26/KWh.

In terms of capacity, the program has seen much higher demand on the commercial side. This is highlighted by the fact that applications for commercial installations (~207 MW) constitute 83 percent of the capacity represented by all applications.

The program has seen a good amount of success so far, with ~81 MW of grid-connected PV installed in 2007, and ~59 MW during the first six months of 2008. But this could just have been a huge wave of initial demand. As the generous initial stages of subsidy allocations get exhausted and there is a stepwise drop to the next subsidy levels, the demand could decline due to the cut in the return on investment.

In California, individual cities and counties might also have their own solar PV promotion mechanisms. San Francisco, for instance, has the GoSolarSF program, which provides $3,000 as a onetime incentive for basic residential installations (up to $6,000 for certain installations), and a $1.50/W (up to a maximum of $10,000) incentive for businesses and nonprofits.

In a presentation in June 2008 (Asia Clean Energy Forum), Ted Flanigan, the CEO of the U.S. based EcoMotion, described several disadvantages and limitations of the California Solar Initiative (http://www.adb.org/Documents/events/2008/ACEF/Session5-Flanigan.pdf, September 15, 2008). Here are three:

- It does not provide for market-linked degression.
- The net-metering-based system tends to limit production to on-site uses.
- The initiative is hampered by numerous administrative and bureaucratic procedures, which tend to act as a dampener to the promotion of solar photovoltaics.

The Role of Utilities

Let's round off this discussion with illustrations of the role utilities in the United States are playing in the solar industry. As one would expect, California-based utilities have a significant presence in these rankings.

The Solar Electric Power Association (SEPA) conducted a study on the status of different utilities stand with regard to solar power, and concluded that Southern California Edison (SCE) and Pacific Gas and Electric Company (PG&E) are currently the two major utilities in the United States in terms of solar use.

The study found that SCE leads in total solar electric capacity (MW) and solar capacity per customer (MW/customer). Its ranking is driven by its contracts with the Solar Electric Generating System (or SEGS, total rated capacity: 354 MW), a set of solar thermal power plants based on parabolic trough collectors, which helps SCE meet its peak demand requirements. PG&E leads in the total solar power systems installed for individuals as well as MW/customer.

Another type of classification for the utilities is "investor owned" and "public power utilities." Among investor-owned utilities, Southern California Edison again leads the rankings, while among public power utilities, the Los Angeles Department of Water and Power leads in total solar capacity and the number of customer-located solar power systems.

INTERPRETING NEWS OF INCENTIVE
PROGRAM CHANGES

Although throughout this chapter we have emphasized the importance of incentives and how they affect the solar PV market of a particular place, it is important in interpreting any news of incentive program changes to keep in mind the culture and the bureaucracy inherent to a particular place or market. There are times when certain new incentives or changes in the existing incentives may give a deceptive picture if not analyzed with the existing culture and bureaucracy in mind.

For example, a proposal in Greece that includes removing the existing cap of 790 MW will get a positive response from the market. But the real picture will be very different since the bureaucracy within the program remains unchanged. Earlier in 2006 an enhanced feed-in tariff for PV ranging from 0.40 to 0.50 euro/KWh had been introduced with an additional 40 percent grant for commercial investors that led to 8,000 applications filed with the Regulatory Authority for Energy (RAE). But due to the bureaucracy, there were no significant new installations as a result.

The part the bureaucracy plays in the new program is evident from the fact that the program administrator has not defined the methodology to be followed to release the application approvals. Also, it is expected that the government will not provide a timetable for systems more than 3GW that are currently waiting in line for approval. This means that the government will still unilaterally control the volume of approvals according to undefined criteria.

The bottom line is that, in order to gain from an investment, incentive programs need to be evaluated carefully, keeping in mind the culture and the bureaucratic structure in that particular market.

The incentive programs described above, as well as similar initiatives in South Korea, Japan, China, and India (each of which is described in detail in the appendix), have done a great deal to make solar PV more cost competitive with traditional energy. But the ultimate goal of widespread adoption of renewable energies such as solar cannot be achieved through incentive programs alone. Renewables Portfolio Standards, to which we will now turn our attention, are another critical component to the future of renewable energy.

RENEWABLE PORTFOLIO STANDARDS

Renewable Portfolio Standards (RPS) create a fixed target for renewable energy supply, apply those targets to the retail electricity suppliers, and encourage competition among them to meet those targets in the most efficient manner. The focus of most RPS activity in the United States has been within the states, with activity at that level growing since the late 1990s. These state efforts, rather than those at the federal level, make more sense since the level of insolation is very different at different locations. Local considerations will remain critical even if a national RPS law is passed.

The Current Scenario

A total of 26 states have RPS policies (Figure 2.1), out of which Illinois, New Hampshire, North Carolina, and Oregon implemented policies in 2007.

The four new programs that were included in 2007, and many other RPS policies, were created by legislative action. The RPS programs are mostly implemented by state utility regulatory agencies like public utilities commissions (PUCs) and public service commissions (PSCs).

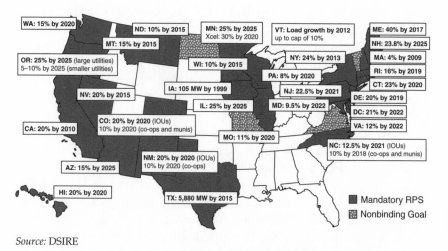

Source: DSIRE

Figure 2.1 State RPS policies and nonbinding renewable energy goals

There are other types of policies with regard to renewable energy, such as:

- Apart from the mandatory RPS policies, some states also have nonbinding energy goals. As the name suggests, these goals are not obligatory in any way. Missouri, North Dakota, and Virginia have such nonbinding renewable energy goals.
- Another scenario is states with nonbinding energy goals changing the policy to mandatory RPS programs. Examples are Illinois and Maine, which previously had nonbinding renewable energy goals but have now changed them to mandatory RPS programs.
- Certain states have developed nonbinding renewable energy goals, in addition to mandatory RPS policies. In such cases, the nonbinding goals are more aggressive. For instance, California has a 33 percent renewable energy goal by 2020.

There is a tendency for states, and government everywhere for that matter, to revise their RPS policies. Such revisions generally strengthen the existing RPS requirements by increasing the renewable energy goals, removing supplier exemptions existing in the policies, or adding resource specific goals. However, policies supporting renewables may be just as easily supported—as we have seen in Spain in 2008, for example—creating an ever-present risk to investing in solar stocks. In 2007, 11 states made substantial modifications in their RPS policies (see Figure 2.2).

Design of RPS in Different States

The design of RPS varies for different states, but the basic premise remains, which is that the retail electricity suppliers or load-serving entities (LSEs) have to procure a fixed minimum quantity of renewable energy. There are major variations in the RPS designs among the states. The variations can be in the renewable energy targets, the timelines, or the renewable energy technologies eligible. Some states have tiered goals comprising separate targets for different renewable resources, along with different schedules and compliance structures. Certain RPS policies have some "preferred" resources for various reasons other than the costs involved.

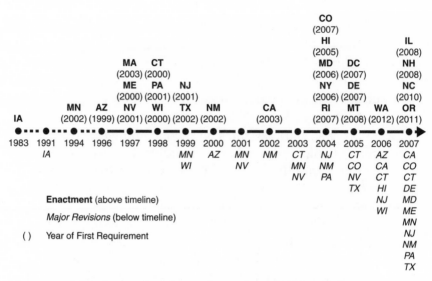

Figure 2.2 Adoption and revision of mandatory RPS policies

Solar Specific RPS Designs

Previously only the least costly projects were considered eligible for RPS programs, but now that trend is changing. Many states have started designing their RPS policies to include solar PV, which has great potential but currently is among the more costly renewable technologies. In order to include such renewable technologies, the RPS provides them with some support. Usually this support comes in the form of credit multipliers, which basically means that the favored renewable technologies are given more credit toward meeting the RPS requirements than other technologies; or as "set-asides," which means that a certain portion of the RPS must be met with a particular technology.

Credit multipliers and set-asides are very commonly used for solar PV. According to the Lawrence Berkeley National Laboratory, set-asides for solar PV currently exist in 12 out of the 26 U.S. RPS programs, as depicted in Table 2.4.

Four states out of the twelve have a combination of both credit multipliers and some form of set-aside. Only two states, Texas and Washington, have credit multipliers without a mandatory set-aside. In 2007 set-asides were created in Delaware, Maryland, New Hampshire, New Mexico, and North Carolina. Also, an expansion was made in the previously created set-aside in Colorado.

TABLE 2.4

TABLE 2.4

Design Elements of Solar Set-Asides

State	First Compliance Year	Resource Eligibility			
		Photovoltaics	Solar Thermal Electric	Solar Heating and/or Cooling	Non-PV Dist. Generation
Arizona	2001	•	•	•	•
Colorado	2007	•	•		
Delaware	2008	•			
Maryland	2008	•	•		
Nevada	2003	•	•	•	
New Hampshire	2010	•	•		
New Jersey	2004	•			
New Mexico	2011	•	•		•
New York	2006	•	•		•
North Carolina	2010	•	•	•	
Pennsylvania	2006	•			
Washington, D.C.	2007	•	•		

Source: Lawrence Berkeley National Laboratory

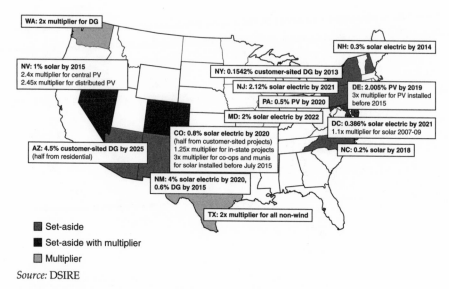

Source: DSIRE

Figure 2.3 Map of credit multiplier and solar set-asides

Figure 2.3 illustrates the credit multiplier and set-aside guidelines per state.

Solar Capacity Required to Meet Existing Set-Asides

According to the Lawrence Berkeley National Laboratory, the following are the largest set-aside-driven solar markets, in terms of required capacity:

- Arizona
- New Jersey
- Maryland
- Pennsylvania

In the next few years significant growth in solar capacity will also be required in:

- New Mexico
- Nevada
- Colorado

Solar generation, as a proportion of expected 2025 statewide load, may be high in many U.S. states, assuming that full compliance is achieved. The Lawrence Berkeley National Laboratory estimates it as:

- 3.1 percent in New Mexico
- 2 percent in Arizona
- 2 percent in Maryland
- 2 percent in New Jersey

Other Differences among RPS Policies: Compliance Frameworks and Eligible Participants

- The compliance framework is a major differentiator among RPS policies in various states. Among those that currently exist, there are three RPS compliance models: In states with retail electric competition, electricity suppliers are usually given broad range to comply with RPS requirements as they see fit.
- In states with regulated utility monopolies, electricity regulators manage the amounts for utility procurement and contracting under the RPS. Again the degree of this management varies from state to state.
- In New York and Illinois, a state agency has the direct responsibility to conduct energy procurements under the RPS.

Another difference is in terms of the entities included under the program. Some of the variations:

- Many states have exempted certain LSEs or end-use customers from meeting RPS requirements. In particular, states often exempt some or all pay own utilities (POUs) from formal RPS obligations, or instead allow POUs to develop their own renewable energy standards.

- Other types of permanent or temporary exemptions have also been adopted. For instance, exemptions for small utilities, large customers, or customers in utility service territories with a rate freeze.
- Certain other clauses, such as force majeure clauses and cost caps, also function as exemptions by reducing the amount of load subject to RPS obligations. Force majeure, a common clause in contracts, essentially frees both parties from liability or obligation when an extraordinary event or circumstance beyond the control of the parties—such as war, strike, riot, crime, or some natural disaster—prevents one or both parties from fulfilling their obligations under the contract.
- Some states have adopted different eligibility criteria related to geographic location and electricity delivery. Simply put, states want to encourage "homegrown" renewable energy projects in order to reap the benefits of any new resource development and economic stimulation resulting from their RPS policy. So if the utilities must acquire their renewable energy from a third party, another limitation can be placed on the distance of renewable projects from the particular state.

Some of the key policy design trends among the states that created or revised RPS programs in 2007 include:

- The aggressiveness of renewable energy targets increased both through revisions to existing programs and through implementation of new RPS policies.
- The use of resource-specific tiered targets increased to a large extent, especially for solar, but also for other favored renewable resource options, such as wind power.
- RPS policies were applied to the POUs also, with three of four new state policies broadly applicable to all electricity suppliers. Revisions made to existing policies also increasingly required POUs to meet renewable energy purchase objectives.

Table 2.5 sums up the key policy design elements for each of the 26 states that have a mandatory RPS programs, as well for the four states with nonbinding goals.

The Impact of Recent Developments on RPS

The purchase agreement of Pacific Gas & Electric (PG&E) with SunPower (250 MW) and Optisolar (550 MW, acquired by First Solar) has put pressure on state and local officials to continue making progress on policy and regulatory initiatives to support solar PV as a means of reaching RPS goals. Another major program in 2008 was the one at Southern California Edison, with its $850 million for solar rooftops.

Although the headline value of PG&E's announcement is significant for SunPower stock trading over the short term, the announcement has a more subtle value that will take years to be realized. The true value to shareholders investing in solar stocks from both SCE's solar rooftop program and PG&E's 800 MW deal is the implied change in business models. The possibility exists that in the next several years utilities own solar generation, opening large markets for solar PV in the United States and unleash value embedded in solar electric generation so far unrealized in the market.

Solar Business Model Transition and RPS

A barrier to deep and volume-driven commodity markets for solar PV in the United States is due in part to the lack of significant involvement of utilities in playing a role in the development of the industry (such as systems ownership). Today's solar PV business models center on user-owned and third-party-owned solar PV systems.

To create large markets for solar PV in the United States, the passive role of utilities must change to accommodate a changing paradigm brought on by the disruption of utilities' existing business models and to better capitalize on solar PV's unique benefits. Indeed, at grid-connected rates of 10 to 15 percent of a utility company's peak-load generation, utilities will be driven by concerns for grid infrastructure (a reduction of traditional capacity additions; reduced substation requirements, and so on); safety (PV systems generate power

TABLE 2.5

RPS Targets and Policy Design Elements

State	First Compliance Year	Current Ultimate Target	Existing Plants Eligible[1]	Set-Asides, Tiers, or Minimums	Credit Multipliers
Mandatory RPS Obligations					
Arizona	2001	15% (2025)	No	Distributed Generation	None[2]
California	2003	20% (2010)	Yes	None	None
Colorado	2007	20% (2020): IOUs 10% (2020): POUs	Yes	Solar	In-State, Solar, Community-Ownership
Connecticut	2000	23% (2020)	Yes	Class I/II Technologies	None
Delaware	2007	20% (2019)	Yes	Solar, New/Existing	Solar, Fuel Cells, Wind
Hawaii	2005	20% (2020)	Yes	Energy Efficiency	None
Illinois	2008	25% (2025)	Yes	Wind	None
Iowa	1999	105 MW (1999)	Yes	None	None
Maine	2000	40% (2017)	Yes	New/Existing	None
Maryland	2006	9.5% (2022)	Yes	Solar, Class I/II Technologies	Wind, Methane
Massachusetts	2003	4% (2009)	No	None	None
Minnesota	2002	25% (2025) 30% (2020) Xcel	Yes	Wind for Xcel Goal for Community-Based Renewables	None
Montana	2008	15% (2015)	No	Community Wind	None
Nevada	2003	20% (2015)	Yes	Solar, Energy Efficiency	PV, DG, EFF, Waste Tire
New Hampshire	2008	23.8% (2025)	Yes	Solar, New, Existing Biomass Methane, Existing Hydro	None
New Jersey	2001	22.5% (2021)	Yes	Solar, Class I/II Technologies	None

State	Year	Target	Existing[1]	Set-Aside or Credit Multiplier[2]	Other
New Mexico	2006	20% (2020): IOUs 10% (2020) Co-ops	Yes	Solar, Wind, Geothermal or Biomass Distributed Generation	None[3]
New York	2006	24% (2013)	Yes	Distributed Generation	None
North Carolina	2010	12.5% (2021) IOUs 10% (2018) POUs	Yes	Solar, Swine Waste, Poultry Waste Energy Efficiency	None
Oregon	2011	25% (2025) Large 5–10% (2025) Small	No[4]	Goal for Community-Based and Small-Scale Renewables	None
Pennsylvania	2001	8% (2020)	Yes	Solar	None
Rhode Island	2007	16% (2019)	Yes	New/Existing	None
Texas	2002	5,880 MW (2015)	Yes	Goal for Nonwind	All Nonwind
Washington	2012	15% (2020)	No	None	Distributed Generation
Washington, DC	2007	11% (2022)	Yes	Solar, Class I/II Technologies	Wind, Solar, Methane
Wisconsin	2000	10% (2015)[5]	Yes	None	None
Nonbinding Renewable Energy Goals[6]					
Missouri	2012	11% (2020)	Yes	None	PSC Authorized to Do So
North Dakota	2015	10% (2015)	Yes	None	None
Vermont	2006	Up to 10% (2012)[6]	No	None	None
Virginia	2010	12% (2022)	Yes	None	Wind, Solar

[1] Some RPS policies allow existing facilities, but only if built after a certain date, e.g., 1995 or 1999. For the purpose of this table these states are identified as not allowing existing resources, because they do not allow older existing facilities. In other states, such as Texas, existing facilities may qualify toward the RPS, but with restrictions not identified in the table. Note also that even those states that do not broadly allow existing facilities to qualify under the RPS offer allow incremental generation from such facilities to qualify.

[2] Credit multipliers were once used extensively but are now being phased out and replaced by set-asides.

[3] Only plants placed in service on or after January 1, 1995, are broadly eligible, except that certain small hydro facilities owned by Oregon utilities and placed in service prior to 1995 are also eligible (such facilities must be certified as "low impact", however, and there are limits to the amount of hydro generation that is allowed to qualify). Incremental efficiency and capacity upgrades on pre-1995 renewable facilities are also eligible.

[4] Targets vary by utility, but the statewide goal is 10% by 2015.

[5] Target equals load growth between January 2005 and January 2012, capped at 10% of 2005 load. The target becomes mandatory in 2013 if the nonbinding goal is not achieved. Though not reflected in this report. Vermont also passed legislation in March 2008 establishing a new, nonbinding goal that 20% of statewide electricity sales be derived from renewable generation by 2017.

[6] In addition to the four nonbinding state renewable energy goals noted here, California, Iowa, and Texas have both mandatory RPS policies and more aggressive nonbinding goals.

back into the grid, thereby creating sources of hazard in outages for linemen and other utility workers); and revenue dilution.

Current business models are costly and complex. Utility involvement may eventually simplify and promote efficient deployment of large volumes of solar PV but not for a number of years to come. The PG&E announcement is one such step in the right direction.

Solar PV market penetration relies heavily on public policy support, technology, and utility support. As policy and regulatory support and technology advance utility involvement in the market, the utility companies represent more than just new sources of large demand. Utility-owned solar PV will unlock additional value not clearly evident today.

By many estimates the cost reduction opportunity in module distribution, design, installation, ownership, and maintenance may be substantially reduced by next-generation business models that fully integrate solar PV systems with module supply in a utility owned model. Such value-enhancing opportunities rely on more regulatory support for PV, such as uncapped net-metering, standardized and simplified inexpensive national interconnection standards, and performance-based rate-making. These regulatory hurdles remain high, and it is not wise for investors to discount wholesale changes in the next few years. Other regulatory changes will also take time. Both revenue decoupling to encourage energy efficiency and conservation, as well as tariff structures that optimize solar PV generation characteristics, figure greatly into changes in regulation that currently free up mostly unrealized value from solar PV.

Investors must also recognize that policy driven markets require further development of Renewable Portfolio Standards in the United States, especially those with solar PV "carve-outs." At a federal level, greenhouse gas emission reduction programs are required to price emissions more efficiently. Lastly, utilities are waiting for state-level economic development initiatives and the growth of state energy initiatives to provide visibility on returns on investment. Policy changes driving utilities forward are evident; however, we do not expect sufficient momentum and the introduction of new and large amounts of demand until the middle to the end of 2010 and beyond.

There is continuing evidence that utilities are expanding their activities and taking the first steps in becoming involved in solar PV.

SCE has been more progressive than most. The utility has put forth a plan to install solar panels on 150 commercial rooftops over the next five years, creating the largest solar PV-generating operation in the world. The project represents 250 MW, enough to power 160,000 homes. SCE asked the California Public Utilities Commission on March 27, 2008, for approval to commit a total of $875 million to the utility's solar project, informing regulators that the expected capacity cost per installed watt would be approximately $3.50, half the average of other photovoltaic installations. Subsequently, on May 8, SCE provided additional cost projections to regulators, telling them it forecast an energy cost of approximately 20 cents per kilowatt hour after adjusting for time of delivery.

SCE intends to invest 4 percent of its five-year capital budget on its solar program; that represents 10 percent of the total of its renewable energy enabling technology budget. SCE's solar program is likely a first in a move by utility companies to begin changing their business models to own solar PV capacity. SCE has estimated it can earn 12.5 percent ROE, which is 100 b.p. above the CPUC approved rate.

SCE has defined peak capacity as "quick [10 minutes or less] black-start" units and third-party peaking capacity supplied under contract by others. SCE has four 45 MW black-start units (180 MW). However, due to competitive reasons, the utility isn't disclosing its third-party peaking capacity. SCE's peaking requirements in the summer months are fully twice the requirements in the winter months. This is the key to understanding the utility's role as a first-mover. Additional reasons why California is home to the first-mover utilities is because there is a strong RPS, good policy support, and a willingness of the state's PUC to grant cost recovery to solar projects. That said, as noted earlier, both the SCE and PG&E initiatives do validate a shifting landscape in the U.S. utility market.

U.S. INVESTMENT TAX CREDIT AND THE IMPLICATIONS OR THE AMERICAN REINVESTMENT AND RECOVERY ACT

Recently, the U.S. Senate passed a bill for renewable energy that grants more than $17 billion as tax incentives. This is a very good boost for the solar industry, considering that it is still in its nascent stage.

This energy legislation:

- Extends the Investment Tax Credit (ITC) for eight years through 2016 and provides tax credits for residential and commercial solar systems
- Gives a 30 percent tax credit to homeowners who install solar systems and businesses that install solar, wind, geothermal, and ocean energy systems
- Gives a 10 percent tax credit to homeowners for energy-efficiency improvements, such as insulation, replacement windows, water heaters, and heating and cooling equipment
- Offers a tax credit of $2,500 to $7,500 for plug-in electric cars, depending on the battery capacity of the vehicle

The bill also lifts the earlier cap on a residential federal income tax credit homeowners could claim for solar PV system installations. Earlier, homeowners could deduct up to 30 percent of the cost, up to $2,000. Solar installations typically cost more than $20,000, and the new law would lift the $2,000 cap. The IRS estimates that nearly 26,000 taxpayers filed for the credit in 2006.

Also, according to this bill the electricity generation companies would be required to generate at least 15 percent of their total electricity from renewable resources by the year 2020.

Implication 1: Strengthening the
U.S. Solar PV Market

One of the implications of this bill is a long-term commitment by the government that will help the United States strengthen its solar PV market. Companies such as SunPower, Evergreen Solar, and Energy Conversion Devices are expected to benefit, leading to a stronger solar PV market in the United States, and therefore globally. But this cushion may not act fast enough to be able to manage the price declines in the year 2009. However, from 2010 on, the United States can be expected to develop into a key solar PV market player, as the impact of higher electricity costs and stronger utility involvement come into play.

Below are some of the U.S. based stocks that showed major one-day gains immediately after the bill was passed, on September 24, 2008:

- First Solar, Inc., was up almost 5 percent at $221.8
- SunPower Corp. was up almost 7 percent at $93.3
- Suntech Power was up over 4 percent at $42.22
- Evergreen Solar Inc. was up over 11 percent at $6.2
- Energy Conversion Devices Inc. was up over 13 percent at $63.79

Because of delays in implementing these programs and continued contraction of credit, all of these shares were trading significantly lower at the time of publication.

Implication 2: New Jobs in the U.S. Solar PV Market

According to an SEIA-commissioned study by Navigant Consulting, 440,000 permanent jobs will be created and $232 billion will be invested by 2016 with this eight-year extension of the ITC.

Implication 3: Boost to the Utilities

The bill also gives utilities access to the solar investment tax credit, which was never given before this. This addition of utilities is expected to boost many more large scale installations, which will further drive the demand for solar PV.

Implication 4: Hit on Oil and Gas Companies

Another implication of this bill is that the oil and gas companies will have to take the hit for these renewable energy sources tax credits, as they will have to pay higher taxes. The revenue-generating provisions of the bill include cutting down on a tax break oil companies get for job creation and overseas production.

NOTE

1. http://www.cpuc.ca.gov/PUC/energy/electric/RenewableEnergy/feedintariffs.htm.

CHAPTER 3

Comparative Business Models: PV Economics and Emerging Markets

Investing in solar stocks can include investing globally across a wide range of companies. Although the end market for solar products is a global commodity market, prices and products are not readily fungible across geographies, nor are prices homogeneous for all players. The solar market of the past several years, constrained by the availability of polysilicon supply, has been punctuated by large differences in prices throughout the supply chain and driven by national incentive programs and differences in electricity grid standards. As competitive forces evolve, investors must understand these dynamics and apply them to both short-term investment strategies and long-term investment valuation calculations that we will discuss in later chapters.

Every company today has specific fundamental factors that determine its short-term margin profile: silicon raw material strategy (raw material costs), manufacturing costs, capacity utilization rates, breakage yields, technical differentiators, and vertical integration profile. Each of these factors plays a critical role in the success or failure of a stock's price performance in the short run. Although we will cover each factor in turn throughout the book, an investor's greatest advantage comes from a thorough understanding of what matters in the long run: the solar value chain. A company's place in this value chain, and how that company might better itself to capture margins as profits shift between segments over time and as end markets evolve, is critical to the long-term valuation of the company.

THE SOLAR SYSTEM AT A GLANCE

Solar companies can be thought of as coming from either industrialized nations or emerging ones. From a strategic standpoint, solar companies in the developed nations and those in emerging economies are virtually indistinguishable. For example, many established European solar cell manufacturers are pursuing economies of scale in their manufacturing processes by using automated equipment to minimize labor costs and maximize quality control as they grow. The investment in new equipment gives them higher operating cost structures when compared to companies in emerging economies.

In the emerging economies of Southeast Asia, China, and Taiwan in particular, solar companies operate under a corporate strategy that utilizes labor to perform tasks handled by equipment in the West. There are investors in solar that are betting on the long-term success of this strategy, based on the view that its benefits are going to emerge as the decisive factor in one to two years. These investors believe that as the cost of polysilicon supplies for Chinese and Taiwanese companies comes down, the relatively inexpensive cost of manual labor in those countries will translate into superior operating cost structures and higher margins. This, in turn, will produce higher valuations for emerging-market companies, and lower relative valuations for European and U.S. companies. Although this rationale seems compelling, it is potentially a faulty long-term strategy.

It's true that the gross margins of most Asian solar producers are deceptively depressed by their comparative lack of access to silicon supplies and that their overall operating costs are lower. But investors who believe that the Asian producers' current lower operating cost structure will be maintained may be disappointed. Given how critical operating cost structure is to long-term profit margins, investors should realize that as solar companies in emerging economies scale up factory sizes to industrial counterparts, their manufacturing strategy becomes essentially the same one we find in the developed world. As companies embark on building 500- and 1,000-megawatt factories, labor intensive manufacturing will be pressed to the brink. Indeed, interviews at Suntech, China's largest cell and module company, revealed that above 500 megawatts of production it becomes suboptimal to manage operations and logistics. Quality control would require too many layers of supervision to be considered effective.

As the solar industry in both the developed and emerging economies grows, it is unlikely that labor intensive manufacturing will be able to keep up with technology advances and factory scale. In the coming years, operating cost structures will equalize globally, throwing open the door to price competition based on pure commodity economics.

VERTICAL INTEGRATION IN THE SOLAR PV INDUSTRY

Vertical integration in the solar PV industry can only be discussed in the context of the polysilicon shortage of 2007–2008. Essentially, vertical integration can occur for many reasons. For the purpose of illustration, let us consider the case of Renewable Energy Corporation of Norway. This company has stated the following key drivers behind its vertical integration strategy (Source: Renewable Energy Corporation presentation, Piper Jaffray conference, June 25, 2008, slide 3):

- Better utilization of capacity
- Better control over growth
- More opportunities to cut costs
- Allows the company to be positioned in the long term across all aspects of the value chain, acting as a "strategic hedge"

In the solar industry, perhaps the most important driver of vertical integration has been the desire by solar companies to access polysilicon either in raw form or in the form of silicon wafers. Companies range from fully integrated, like Renewable Energy Corporation, to nonintegrated, like Solarfon. Those in the first category have an advantage and should benefit from margins throughout the supply chain. Therefore, "theoretically" they should net a higher profit and receive a higher stock valuation.

POSITION IN SUPPLY CHAIN AND EXTENT
OF VERTICAL INTEGRATION

Different companies in the solar industry occupy one or more positions in the supply chain. While some companies focus their manufacturing on PV cells, modules, and panels, starting from silicon wafers, others are integrated upstream with in-house facilities for producing silicon ingots and wafers. For instance, JA Solar Holdings manufactures cells in-house and the cells are its end product, whereas SunPower manufactures cells, modules, and panels in-house, and its end products are modules. Also, these companies try to capitalize on geographical advantages; for example, SunPower manufactures the cells and panels in the Philippines. Table 3.1 lists the biggest cell and module companies in the market as of the end of 2008.

Investors who want to invest in companies based on their supply chain characteristics should find Tables 3.2 and 3.3 helpful.

So, we observe that Mc-Si technology is most widely used; companies using Mc-Si include JA Solar Holdings Co Ltd, SunPower Corporation, Trina Solar Limited, Canadian Solar Inc., China Sunergy Co Ltd., Suntech Power Holdings, LDK Solar Co Ltd., and Renesola. Evergreen Solar makes use of string ribbon technology to

T A B L E 3.1

Largest Cell and Module Companies, Year-End 2008

Company	2007	Q108A	Q208A	Q308E	Q408E	CY08E	CY09E
Suntech	363.0	110.2	111.5	138.0	130.3	490.0	800.0
Canadian Solar	78.4	40.0	46.1	59.5	90.9	236.5	529.0
First Solar	199.9	79.0	103.2	140.5	189.5	512.2	1,090.0
Solarfun	78.3	40.3	43.1	47.7	51.3	182.4	275.3
REC	42.0	13.0	20.0	18.0	34.0	85.0	220.0
SunPower	101.5	38.5	49.8	65.8	82.0	236.1	460.0
Trina Solar	75.9	29.5	47.6	66.4	60.0	203.4	351.0
Yingli Green	142.6	54.6	68.2	72.9	84.6	280.3	533.8
JA Solar	113.5	48.9	51.7	89.1	76.0	265.7	729.0
Q-Cells	389.2	117.0	146.5	153.5	183.0	600.0	1,000.0
CSUN	74.0	22.6	32.8	36.0	43.6	135.0	235.0

Source: Alternative Energy Investing, LLC

T A B L E 3.2

Positions of Crystalline-Silicon Solar Companies in the Value Chain

	Polysilicon	Ingot	Wafer	Cell	Module	Installation
Yingli Green		■	■	■	■	■
SunPower				■	■	■
Trina Solar		■	■	■	■	
Canadian Solar		■	■	■	■	
Solarfun Power		■	■	■	■	
Evergreen Solar			■	■	■	
Suntech Power				■	■	
JA Solar				■		
China Sunergy				■		
LDK Solar	■	■	■			
Renesola	■	■	■			

Source: Alternative Energy Investing™

manufacture wafers. The solar companies also have different positions in the solar PV value chain.

INDUSTRY ANALYSIS OF THE SOLAR SUPPLY CHAIN

The "attractiveness" of each segment of the solar supply chain using Porter's "Five Forces" analysis gives an investor further input to supplement her valuation of companies in each part of the chain.

Before we explain this in the context of the solar PV industry, a brief discussion of Porter's Five Forces" is in order. Without getting into too much detail, Porter's analysis is a way of identifying how attractive an industry is in terms of potential returns. The Five Forces, in simple terms, are:

1. *Threat from New Entrants.* If entry barriers are high (due to technology access, capital costs, economies of scale, etc.), an industry becomes more attractive.

TABLE 3.3

Different Technologies of Companies in Solar Supply Chain

Company Name	End Product	Technology	What Is Manufactured In-House	Comments
JA Solar Holdings Co Ltd	Cells	Mc-Si	Cells (outsource wafer supply)	
SunPower Corp	Modules and systems installation services	Mc-Si	Cells, modules, panels, also integrated upstream into installation	Cells and panels made in Philippines; Korean ingot/wafer JV
Trina Solar Ltd	Modules	Mc-Si/Pc-Si	Ingots, wafers, cells, modules	Source poly and reclaimed Si externally
Canadian Solar Inc	Modules	Mc-Si/Pc-Si	Cells, ingot/wafer modules (manufacturing in China)	Also source cells externally
China Sunergy Co Ltd	Cells	Mc-Si/Pc-Si	Cells	Purchase wafers
Suntech Power Holdings	Modules	Mc-Si/Pc-Si	Cells, modules	
Yingli Green Energy Hold	Modules	Pc-Si	Pc-Si ingots, wafers, cells, modules, systems, installation	High degree of vertical integration
Evergreen Solar Inc	Modules	Pc-Si	Wafers, cells, modules, panels	Wafers produced with "string ribbon" technology
Solarfun Power Hold	Modules	Pc-Si	Cells, ingot/wafer modules	Also purchase cells to supplement their own production
Energy Conversion Devices	Modules	a-Si	TF modules; also design and manufacture their own production equipment	Source stainless steel, argon, nitrogen, silane, germane, etc.
First Solar Inc	Modules	CdTe	TF modules	
LDK Solar Co Ltd	Wafers	Mc-Si/Pc-Si	Ingots, wafers, integrating into polysilicon production (7,000 tons planned by end of 2008)	Source polysilicon
Renesola Ltd	Wafers	Mc-Si/Pc-Si	Ingots, wafers, polysilicon	

Source: Alternative Energy Investing™

2. *Threat from Existing Rivals.* The more fragmented an industry, the more the competitive intensity and rivalry in the industry and the less potential for good returns.

3. *Threat from Substitutes.* If the costs of switching are relatively low and there are substitute products involving reasonable trade-offs, the potential for good returns decreases, and the industry becomes less attractive.

4. *Bargaining Power of Suppliers.* If an industry has very few suppliers—for instance, if the supplier base is highly concentrated—the bargaining power of companies in the industry decreases, and the industry becomes less attractive. However, if the supplier base is highly fragmented, companies in the industry would be able to negotiate better terms with suppliers, hence increasing the potential for returns.

5. *Bargaining Power of Buyers.* If the customer base for an industry is highly concentrated, bargaining power for companies in the industry decreases, and the industry becomes less attractive. On the other hand, if the customer base for an industry is highly fragmented, then bargaining power for companies increases, and the industry becomes more attractive in terms of potential returns.

Applying the Five Forces model to the solar supply chain yields the following analysis.

Silicon: Poly Ingots/Wafers

The industry is currently concentrated, with a few large players bringing in most of the capacity. However, in the coming few years the degree of concentration is poised to decrease, with new players coming in.

For several years polysilicon has not had "substitutes" for the manufacturing of crystalline silicon-based PV. More recently, upgraded metallurgical silicon has emerged as a viable substitute, with reputable companies such as Q-Cells deciding in favor of it. As and when confidence in upgraded metallurgical silicon increases, this could emerge as a viable substitute for the traditional polysilicon route, thus driving down the attractiveness of polysilicon companies.

The polysilicon part of the supply chain is characterized by high capital costs, long lead time for setting up capacity, economies of scale, and difficulties in mastering the intricacies of the manufacturing process. These create high barriers to entry.

The silicon raw material for poly manufacturing does not have too many supply constraints, so the bargaining power of suppliers is low.

With so many cell/module companies emerging in the last few years, and capacity constraints in polysilicon, the bargaining power of buyers has been relatively limited. Moreover, although some cell/module companies have attempted to backward integrate into polysilicon, the level of difficulty involved in manufacturing implies that not all companies have been willing or even capable of adopting this route.

The high level of attractiveness of the silicon industry in recent years is reflected in the high prices for poly over the last few years. Another area where this is reflected is the fact that cell/module companies have been willing to invest large amounts of capital as prepayments for polysilicon projects.

Of course, as more capacity emerges, and when confidence in upgraded metallurgical silicon increases sufficiently, the attractiveness of the poly industry will tend to decrease.

Cell/Module Companies

This part of the supply chain is less concentrated than the polysilicon part, which means there is greater rivalry among existing players.

In a narrow sense, given the incentives for renewable sources of electricity, substitutes for cell/module companies are other sources of alternative energy, such as wind energy. Broadly speaking, in the absence of subsidies, substitutes for solar PV cells/modules could include any other sources of electricity generation.

This part of the supply chain is again technology intensive, and there are also economies of scale. Still, several new companies have entered the industry in recent years due to attractive incentives for solar PV in several countries. In years to come, however, one can expect scaled-up players to dominate and entry barriers to increase.

For crystalline silicon-based PV manufacturers, the bargaining power of suppliers (polysilicon manufacturers) is relatively high at present. For thin-film-based PV manufacturers, the bargaining power of suppliers is comparatively lower.

So far, demand for solar PV has remained robust as a result of strong incentives. However, given that solar PV products are mostly commoditized and involve little switching costs between similar projects (project application and location), the bargaining power of buyers is poised to increase in the future.

System Integrators/Installers

This part of the supply chain is the least concentrated compared to the other parts. There are numerous installers for solar PV in every region.

Entry barriers are minimal because there are limited technology/scale constraints, which mean lower barriers to entry.

The bargaining power of suppliers (cell/module companies) has been relatively high during these last few years of tight supply conditions. Given that it is not easy for a system integrator to backward integrate into cell/module manufacturing, the potential for "supernormal" returns for companies in this part of the supply chain appears limited.

The bargaining power of buyers is again on the higher side because customers can choose one system installer over another without seeing too many significant differences in service quality and price.

Given the relatively low level of attractiveness of this part of the supply chain, system integrators need to follow a few unique approaches (in terms of product/service) in order to differentiate themselves from competitors. For example, Akeena Solar has its unique "Andalay" system, which is differentiated from competitors and helps Akeena Solar stand out. Some of the features of the Andalay system are:

- Better reliability as compared to competitors
- Superior aesthetics of the panels
- Strongly backed by warranties

SOLARFUN AND SUNTECH POWER: COMPARISON OF INTEGRATION STRATEGIES

To illustrate why investing in quality companies matters in the solar sector, we will compare the integration strategies of two solar companies, Solarfun and Suntech Power.

Solarfun

Solarfun has been heavily reliant on outside suppliers for much of its raw material requirements. However, Solarfun has been in an unfavorable position as far as its raw material supply is concerned. We can identify three main issues related to this:

1. It has been dependent on just five suppliers for ~60 percent of its raw material requirements, so if these suppliers face problems in fulfilling their commitments, it could have a significant impact on the company.
2. Being a relatively new entrant, it lacks entrenched relationships with reliable suppliers. This is reflected in the fact that there have been instances in the recent past where Solarfun's suppliers—many of whom have a limited track record—have been unable to deliver on their commitments.
3. Most of its supply contracts have been extremely short-term arrangements of less than two years in duration.

Solarfun's response to these critical issues has been to pursue an integration strategy to obtain as much control as possible over the upstream portions of the silicon raw material supply chain. When we analyze this, we find that the quality of Solarfun's integration strategy leaves much to be desired in the eyes of an investor, largely because Solarfun is trying to venture into too many complex areas within a very short time frame.

To understand Solarfun's integration strategy, it would be useful to analyze the different parts of the raw material chain.

Modules/Cells

Solarfun had a target of 160 to 180 MW of module shipments in 2008, with plans to expand this in the future as cell manufacturing capacity grows. All of Solarfun's modules are produced using its own cells. Solarfun has four Mc-Si cell lines and four Pc-Si cell lines, each with 30 MW of capacity, which implies the aggregate annual cell capacity is 240 MW. This was expected to reach 420 MW by 2009, but is currently being reevaluated due to credit conditions.

Wafers

As this book was being written, Solarfun was installing six wire saws for wafering. The company also plans to purchase 54 additional wire saws to increase annual wafer capacity to 300 MW. This is expected to be completed by March 2009. Solarfun also has the following wafer supply arrangements with external suppliers:

- An agreement with LDK to secure Pc-Si wafers worth ~$285 million from early 2008 to 2010
- Contracts for wafers worth $230 million over a seven-year period with a Korean conglomerate, with volumes >30MW/year from 2011
- A 140 MW long-term wafer contract with Wacker Schott Solar GmbH (a JV between Wacker and Schott Solar) signed in January 2008

As one would expect, we see here that Solarfun's plans for wafer supply are more or less in line with its planned cell capacity expansions, with a focus on eventually being able to source all of its wafer needs internally.

However, we need to note here that wafering is an area that is almost completely new to Solarfun. At this stage the company does not have any experience of note to be able to assure an investor that its wafer facilities would operate as required. Any failure in Solarfun's wafer operations would adversely (and directly) impact its cell and module manufacturing operations.

Ingots

As of 1Q 2008, only 10 percent of Solarfun's ingot requirement was produced internally. However, this is set to increase with the recent acquisition of 100 percent of Yangguang Solar, which gives it access to 50 to 60 MW of ingot capacity in 2008. Yangguang is expected to expand capacity to 200 MW by year-end 2008 and 300 MW by year-end 2009. Besides this, Solarfun has also signed an agreement with a supplier to deliver 178 MW of ingots and wafers for seven years, starting from July 2007.

Here again, Solarfun's focus is clearly on securing all ingot supplies internally as it expands cell and module capacity.

However, there are several serious problems that could undermine Solarfun's quest for self-sufficiency in ingots. Yangguang is a new entrant in the ingot manufacturing business, having started operations only in the latter half of 2007. Given the complexity of the ingot manufacturing process, and knowing that Solarfun's experience in this field is also limited, it is hard to imagine a completely smooth ramp-up on the large scale and tight time frame the company has planned. Further, serious issues could arise due to any failures on the part of capital equipment suppliers for the ingot facilities.

If Solarfun is unable to meet its ingot targets as desired, it would have a highly detrimental ripple effect on the downstream sections of the supply chain.

Polysilicon
We now highlight some of Solarfun's key supplier relationships vis-à-vis polysilicon.

Solarfun has signed an agreement with GCL Silicon Technology to obtain aggregate supplies of ~1.2 GW of virgin polysilicon over eight years (starting June 2008). This would mean an average supply of ~150 MW every year for eight years. It also has an agreement with Hoku Scientific for $384 million worth of polysilicon beginning (latest) 2009 and continuing over 10 years. Further, it has a contract with E-mei to secure 500 tons of silicon products for five years, starting from the completion of the facility (expected some time in 2009).

As with wafers and ingots, even on the polysilicon front Solarfun could run into a whole host of problems. For example, Yangguang (the in-house ingot facility) is highly dependent on GCL for polysilicon. There are valid reasons to doubt GCL's ability to deliver on its future commitments, given that it has fallen well short of its contractual obligations in the past. If the polysilicon market continues to be tight, Solarfun could be hampered in its attempts to make up for any shortfall through spot purchases.

Conclusion

What we see here is a company that is almost completely new to the PV cell/module business, trying to ramp up its cell/module operations on a large scale as it simultaneously attempts to also integrate downstream (again, without much of a track record and on a large scale), in a business where each and every part of the supply chain involves a mastery of complex technology and manufacturing expertise. In terms of the quality of its integration strategy, this does not present a picture of confidence to an investor.

Suntech Power: Integration Strategy

Suntech Power's quality can be clearly seen in the integration strategy it has pursued. Its focus is on cell and module manufacturing, while establishing a robust supplier network composed of high-, medium-, and low probability new entrant suppliers, combined with short- and long-term commitments to well-established incumbents, to provide silicon inputs. While not integrated upstream in terms of the silicon supply chain, Suntech Power's integration strategy pursues downstream opportunities in the BIPV space and control over process technology and factory equipment.

Modules/Cells

Suntech Power's annual cell manufacturing capacity was one GW by the end of 2008, up from 540 MW at the end of 2007.

Wafers, Ingots, and Polysilicon

Suntech Power depends on external suppliers for wafers, ingots, and polysilicon The company sources its polysilicon and wafers from over 40 suppliers through long-term agreements and obtains a small portion of its requirements from the spot market. This means it has a diversified supplier base plus good supply visibility. One potentially negative point to note here is that in 2007 as much as 45 percent of its silicon (wafers and poly) supplies were met by its top five suppliers.

The company's supplier list includes established players with robust track records. For example, MEMC (with which Suntech Power has a 10-year contract) and Deutsche Solar AG are two of its key suppliers. To bolster the supply situation further, Suntech Power has acquired strategic stakes in poly suppliers such as Hoku Scientific.[1]

During 2008, Suntech Power began moving aggressively to augment its raw material supplies for the long term. On March 13, 2008, it acquired a minority stake in Russian polysilicon producer Nitol Solar for $100 million. On May 27, 2008, the company announced that it had picked up a stake in a China-based wafer manufacturer, Shunda Holdings, for nearly $99 million; there was also a 12-year contract for a total of nearly seven GW of wafers from Shunda starting 2008. In June 2008 it announced news of a long-term contract with Wacker Schott Solar GmbH to procure a total of 220 MW of wafers, and in July 2008, Suntech signed an agreement to procure 260 MW of wafers from PV Crystalox Solar, a British wafer manufacturer. The agreement provided for the supply of wafers over a five-year period from 2008 to 2013, at fixed prices and volumes.

A Push into Downstream Integration

Instead of integrating upstream into wafers/ingots/poly, Suntech has focused on the downstream side, in the form of value-added building-integrated photovoltaics (BIPV) and system integration. As part of its push into BIPV, Suntech has taken a strategic stake in MSK, a company that specializes in this area. Although the push into BIPV makes sense in the long run should solar PV become synonymous with building materials more generally, Suntech did not integrate MSK easily, raising the issue of integration risk for the company's investors.

Focus on Process Equipment and Technology

Since Suntech has established agreements with a wide variety of raw material suppliers and because its key suppliers are well established in the industry, its vertical integration model focuses less on the raw material side and more on process equipment and associated technology.

Its ability to design its own equipment saves capital costs and also permits it to exercise greater control over its processes. One step Suntech has taken in this regard is the acquisition of the KSL-Kuttler

Automation Systems, a German solar capital equipment company. Further, Suntech Power has encouraged its capital equipment suppliers to set up their manufacturing facilities in the vicinity of its plant in Wuxi to facilitate better control over its equipment supplies.

Conclusion

Suntech Power, in contrast with Solarfun, has pursued an integration strategy that reflects quality in the eyes of the investor; however, the execution has not always been the best. That said, the company's focus is mainly on cell/module manufacturing, with strategic acquisitions driving its downstream push and its attempts to control process technology. An established (and relatively diversified) supplier base reinforces this strategy. An investor would therefore have valid reasons to feel more assured about Suntech Power than about Solarfun.

STRATEGIC PARTNERSHIPS IN THE PV INDUSTRY

Sometimes solar photovoltaic companies may decide to enter into strategic partnerships with other companies in order to gain advantages they otherwise would not have. Several possibilities for this kind of partnership exist:

- A company with access to a unique technology deciding to partner with a mass-scale manufacturing company in order to attain scale
- A company licensing its proprietary know-how to another company or group of companies
- A company taking strategic stakes in another company or group of companies to gain access to a portfolio of technologies
- Partnering with another company for the purpose of marketing and distribution

These tie-ups or partnerships can be structured in several ways; as joint ventures, for example. We note that in some cases the basis for partnerships may be a combination of one or more of the principles outlined above.

Consider the following instances:

- Q-Cells AG (Germany), Evergreen Solar (United States), and Norway-based Renewable Energy Corporation (Norway) have a joint venture called "EverQ" for manufacturing string-ribbon-based photovoltaic products. This venture is meant to leverage the strengths offered by each of these companies in order to maximize the value of the unique string ribbon technology.
- Tata BP Solar, a joint venture between India-based Tata Power and BP Solar, was established more than 15 years ago.

An investor would need to consider the potential value in such strategic stakes (if any) when evaluating a solar PV company. If a solar company has a partnership arrangement with a company (or group of companies) that looks promising, it would make the company's stock worth more.

If a solar PV company has differentiated and valuable technology but finds that the full value of this technology can be realized only through licensing agreements and joint ventures, it may decide in favor of this strategy. Of course, a company may sometimes decide to continue in its primary PV manufacturing business while supplementing this with various joint ventures and/or licensing agreements.

A good example of this is Evergreen Solar, which, as noted above, has the "EverQ" joint venture with Norway's Renewable Energy Corporation (REC) and Germany's Q-Cells AG. EverQ is a strategic move by Evergreen Solar to extract more value from its proprietary string ribbon technology by leveraging the strengths of its two joint venture partners. Indeed, plans to expand EverQ capacity to 600 MW by 2012 is intended to help Evergreen Solar derive more benefits from string ribbon technology than would otherwise be possible.

Other companies with unique technological capabilities may also choose to supplement their primary business operations with the licensing route in an attempt to utilize intellectual property more effectively.

LONG-TERM CONTRACTS FOR SOLAR
MODULES AND CAPACITY EXPANSIONS

In the recent past, especially during the years when silicon shortages plagued the industry, companies engaged in long-term contracts as a growth strategy that was seemingly good at the time. Companies signed long-term contracts, and investors rewarded them with higher valuations. As a consequence, more capital became available for capacity investment. From late 2005 through early 2008, long-term contracts were an effective means of securing scarce materials. Companies could use these contracts to show investors that they could take advantage of market opportunities without any risks of running short. Long-term polysilicon and wafer contracts were the object of much strategic activity. In the frenzy of the shortage, however, long-term contracts for modules emerged as well, take-or-pay contracts that often covered a period of years.

In this section, we present examples of long-term contracts that provided the sales visibility that investors paid for when supplies and devices were in short supply. These examples, however, could become sources of value destruction in the future should the period of surplus result in broken contracts. Indeed, the record for engaging in long-term contracts was yet to be written at the time this book was published. Therefore, the degree to which they may have been a bad thing is not yet known.

Evergreen Solar

On May 22, 2008, Evergreen Solar announced that it had signed a $750 million agreement with German company Ralos Vetriebs GmbH for deliveries from 2008 through 2013.

On June 29, 2008, the company announced that it had signed agreements with two companies, namely groSolar and Wagner & Co Solartechnik GmbH, valued at $1.7 billion and extending through 2012.

On July 15, 2008, Evergreen Solar announced a $1.2 billion long-term sales contract with IBC Solar AG, Germany, extending through 2013.

After the news of the IBC contract broke on July 16, 2008, around 70 percent of the expected production at the company's Devens plant was sold out through 2010, and fully 100 percent of expected production at Devens from 2011 through 2013 was sold out. This kind of visibility is very valuable to an investor because it means the company is substantially insulated from external, demandside risks for the duration of these contracts. The risks to the investor arise instead primarily from execution (ramping capacity, ensuring adequate yields and throughput, and so on), which is internal to the company.

First Solar

On January 9, 2007, First Solar announced that it had amended four existing contracts so 264 MW of additional sales would result between 2009 and 2012. On July 9, 2007, First Solar signed five agreements with customers for an aggregate of 685 megawatts, for a value of ~$1.28 billion, between 2007 and 2012. This announcement included contracts signed with EDF Energies Nouvelles, RIO Energie GmbH, and SunEdison LLC.[2]

As of August 29, 2008, First Solar had contracts in place to supply 3.4 GW to various customers between 2008 and 2012.

Trina Solar Limited

Trina Solar Limited entered into a three-year sales agreement with Ergyca Power Srl, a subsidiary of GreenergyCapital SpA. Under this, the company would supply GreenergyCapital with PV modules at a total value of $158 million. This agreement will give the company long-term growth prospects in the Italian PV market.

Solarfun Power Holdings

Solarfun signed an agreement to supply Q-Cells International with 100 MW of PV modules per year for a period of three years, from 2009 through 2011. The modules would be manufactured and delivered according to Q-Cells' design and specifications and used in the systems business segment of Q-Cells.

LDK Solar

On June 30, 2008, LDK Solar announced a 10-year supply contract with Canadian Solar for an additional 800 MW of modules. Deliveries were scheduled to begin in July 2009, with ~40 MW expected to be shipped in 2009 and ~80 MW from 2010 on.

MEMC Electronic Materials

MEMC Electronic Materials Inc. entered into a $3.5 billion supply deal with Tainergy Tech Co. Ltd., Taiwan, to supply solar wafers for a period of 10 years.

Renewable Energy Corp

On June 25, 2008, news of a contract between Renewable Energy Corp and China Sunergy was announced, under the terms of which $400 million of 156mm wafers would be supplied by REC to China Sunergy. The contract runs from 2009 through 2015.

SunPower Corp

SunPower (through its subsidiary High Plains Ranch II) in California entered into a contract with the California utility Pacific Gas & Electric (PG&E) to supply 250 MW of modules. The project is expected to begin power delivery in 2010 and be fully operational in 2012.

THE ECONOMICS OF SOLAR PV

So far, we've stressed the critical life-giving role of government subsidies and will examine it further here. Until solar electricity is on a par with costs of other sources of traditional electricity available locally, the solar industry will continue to require subsidy support. We consider Europe and the United States separately since different computation methods are involved in each. This is because incentives in Europe are driven by feed-in tariffs, while the United States is driven more by investment tax credits, accelerated depreciation, and other such financial inducements.

Europe

We analyze the economics of solar PV in Europe using two methods:

1. Comparison of yearly income and expense, assuming that the PV system installation is 100 percent based on financing
2. Payback period

Comparing Yearly Income and Expense

A simple way to identify whether the installation of a solar system is economical is to do a cost-benefit analysis, or compare the income versus the expenses.

To calculate income, we first need to know how much electricity gets generated by a solar system each year (Table 3.4). This in turn depends on the insolation (the measure of solar radiation energy received on a given surface at a given time) of the specific place we are considering and varies across different months. We obtain this information using a Web-based energy production calculator provided by the U.S. National Renewable Energy Laboratory.

TABLE 3.4

Average KWh of Electricity Generated

Month	Germany (Frankfurt)	Spain (Madrid)	Italy(Naples)
January	17	76	57
February	45	84	70
March	79	127	92
April	92	124	103
May	100	127	118
June	85	131	121
July	107	143	134
August	101	138	130
September	79	108	107
October	59	97	91
November	19	78	60
December	18	52	54
Year	802	1,285	1,135

Source: PVWatts

This tool, called PVWatts, performs its computations for crystalline silicon-based solar cells based on certain standard assumptions.

Next, we need to know the feed-in tariff per unit of electricity generated. As we explained in Chapter 2, for one kilowatt roof-mounted (partially integrated) installations in 2008, feed-in tariffs are 46.75 euro cents in Germany, ~44.04 euro cents in Spain, and 44 euro cents in Italy.

Based on these tariffs and "KWh generated" estimates, we estimate an annual income of ~€ 375 in Germany (Frankfurt), ~€ 566 in Spain (Madrid), and ~€ 499 in Italy (Naples).

On the "expenses" side, we assume that the PV installation is completely financed. We have two variables, namely, installed cost/W_p and financing cost (percent of total system costs that would have to be paid every year for financing). As can be seen in Table 3.5, we analyze the impact of these variables on annual expenses for a one kilowatt, partially integrated, roof-mounted PV system installed in Frankfurt (Germany), Madrid (Spain), and Naples (Italy).

In Table 3.6 we identify the values of installed cost/W_p and the financing rate for which solar PV would be economically viable for Germany (shaded boxes), where the annual income from the PV system is € 375.

TABLE 3.5

Variation of Annual Financing Costs as a Function of Installed Cost/W_p and Financing Rate

Cost/W_p € 135.00 Financing Rate	€4.00	€4.50	€5.00	€5.50	€6.00	€6.50
5.0 percent	€ 200	€ 225	€ 250	€ 275	€ 300	€ 325
5.5 percent	€ 220	€ 248	€ 275	€ 303	€ 330	€ 358
6.0 percent	€ 240	€ 270	€ 300	€ 330	€ 360	€ 390
6.5 percent	€ 260	€ 293	€ 325	€ 358	€ 390	€ 423
7.0 percent	€ 280	€ 315	€ 350	€ 385	€ 420	€ 455
7.5 percent	€ 300	€ 338	€ 375	€ 413	€ 450	€ 488
8.0 percent	€ 320	€ 360	€ 400	€ 440	€ 480	€ 520
8.5 percent	€ 340	€ 383	€ 425	€ 468	€ 510	€ 553

T A B L E 3.6

Germany (Frankfurt): Identifying Economically Viable Costs/W and Financing Rates

Financing Rate	Cost/W$_p$ €135.00 €4.00	€4.50	€5.00	€5.50	€6.00	€6.50
5.0 percent	€200	€225	€250	€275	€300	€325
5.5 percent	€220	€248	€275	€303	€330	€358
6.0 percent	€240	€270	€300	€330	€360	€390
6.5 percent	€260	€293	€325	€358	€390	€423
7.0 percent	€280	€315	€350	€385	€420	€455
7.5 percent	€300	€338	€375	€413	€450	€488
8.0 percent	€320	€360	€400	€440	€480	€520
8.5 percent	€340	€383	€425	€468	€510	€553

We identify the economically viable (where financing cost is less than or equal to tariff income) installed costs/W and financing rates for Spain and Italy as well, in Tables 3.7 and 3.8 (shaded boxes indicate economic viability).

We find that Germany has the maximum constraints on economic viability despite having the highest feed-in tariffs in the group, while Spain has the least constraints on economic viability. The reason for this is clearly the higher insolation in Spain (1,285 KWh of electricity per year for a one KW installation, compared to just 802 KWh in Germany). This points us to the fact that while feed-in tariffs are an important factor in determining economic viability of solar PV, the extent of insolation that a place receives also matters to a great extent.

Payback Period Method

The concept of "payback period" could also be used to gauge the economics of solar PV. We present these calculations for PV installations in Germany (Frankfurt), Spain (Madrid), and Italy (Naples) in Table 3.9.

The numbers here closely reflect those in the earlier analysis, where we identified economically viable costs/W$_p$ and financing rates for these three locales. For example, if we look at the case where installed costs/W$_p$ are at €4.00/W$_p$, we find that an installation in Madrid would have a payback period of just seven years,

TABLE 3.7

Spain (Madrid): Identifying Economically Viable Costs/W and Financing Rates

Financing Rate	Cost/W$_p$ €135.00 €4.00	€4.50	€5.00	€5.50	€6.00	€6.50
5.0 percent	€200	€225	€250	€275	€300	€325
5.5 percent	€220	€248	€275	€303	€330	€358
6.0 percent	€240	€270	€300	€330	€360	€390
6.5 percent	€260	€293	€325	€358	€390	€423
7.0 percent	€280	€315	€350	€385	€420	€455
7.5 percent	€300	€338	€375	€413	€450	€488
8.0 percent	€320	€360	€400	€440	€480	€520
8.5 percent	€340	€383	€425	€468	€510	€553

TABLE 3.8

Italy (Naples): Identifying Economically Viable Costs/W and Financing Rates

Financing Rate	Cost/W$_p$ €135.00 €4.00	€4.50	€5.00	€5.50	€6.00	€6.50
5.0 percent	€200	€225	€250	€275	€300	€325
5.5 percent	€220	€248	€275	€303	€330	€358
6.0 percent	€240	€270	€300	€330	€360	€390
6.5 percent	€260	€293	€325	€358	€390	€423
7.0 percent	€280	€315	€350	€385	€420	€455
7.5 percent	€300	€338	€375	€413	€450	€488
8.0 percent	€320	€360	€400	€440	€480	€520
8.5 percent	€340	€383	€425	€468	€510	€553

TABLE 3.9

Payback Period (in Years) and Installed Costs/W$_p$

	€4.00	€4.50	€5.00	€5.50	€6.00	€6.50
Frankfurt	11	12	13	15	16	17
Madrid	7	8	9	10	11	11
Naples	8	9	10	11	12	13

while it would be eight years in Naples and eleven in Frankfurt. In fact, for any of the considered costs/W_p, Madrid has the least payback period and Frankfurt the highest. Again, we come to the key point that the degree of insolation in a place is a major determinant of the economics of solar PV.

United States

In the United States, in addition to the federal incentives, there are state incentives for PV that vary widely across different states. Based on the specific incentives available in a particular place, the economics of solar PV in the United States would differ widely from state to state.

For the purpose of analysis, we will consider California, which has set aside $2.6 billion to support the California Solar Initiative (CSI), which has both rebates and feed-in tariffs:

- For residential and small commercial systems less than 100 KW_p for installations in 2007, CSI provides rebates (expected performance based buy-down, or EPBB)
- For larger commercial systems more than 100 KW_p a feed-in tariff is provided at a fixed rate over a five-year period (performance based incentive, or PBI)

Let us first look at the residential and small commercial applications that have solar systems sized less than 100 KW_p. The approximate cost of a three-KW rooftop solar system along with the federal tax credit is approximately $15,000 to $18,000, or $5 to $6 per watt, after deducting all available incentives. The average retail price of residential electricity is approximately $0.13/KWh.

The analysis is dependent on multiple variables, and for our ease we will assume that the solar PV system has:

- Operating life of 25 years
- Discount rate of 4 percent
- Peak sunlight per day of 4.5 hours

A person installing this system would never be able to break even. In order to break even on the solar investment, the installed cost of the solar system would have to be less than $3/watt. This is almost half the current market prices in California.

However, for households that consume electricity heavily, $0.13/KWh is less than the peak rate and is crossed frequently. For such consumers of electricity, installing a solar system could make economic sense, since it might give them positive returns in a reasonable time frame. They could potentially use the solar system to offset the premium rate paid by them, which could at times exceed $0.30/KWh.

In the case of commercial setups in California, there is a complex electricity rate structure. In addition, the incentives or credits given at the state and federal level make it more complicated. The feed-in tariff is limited to 5 years, much less than the 20-plus years of European feed-in tariffs. From an economic perspective, though, the tax incentives are very attractive; they include the federal investment tax credit of 30 percent of the cost of the system, and the accelerated depreciation (five years) to lower near-term taxes. The grid cost of electricity will be lower for commercial and industrial customers, but the tax incentives are much higher.

For large commercial applications that have solar systems greater than 100 KW_p, CSI gives a PBI ("performance based incentive") that pays a fixed feed-in tariff rate for solar PV generated electricity. Similar to the rebate program, the PBI is on a sliding scale based on installed MW_p milestones. The PBI plus tax incentives (ITC plus accelerated depreciation) can make the large commercial systems economically attractive. The federal 30 percent investment tax credit (ITC) offers a convincing complement to system economics. Although it is not as generous as many of the European feed-in tariffs, the ITC makes the solar systems economically more viable. The ITC was renewed in October 2008.

For analyzing the economics of a PV system, another factor to be considered is the cost of the grid supplied electricity. The value of the electricity produced by the solar PV installation is actually the cost of the electricity provided by grid that is being replaced. The electricity bill would be higher in the summer season, as the electricity consumption increases sharply; fortunately, the solar PV system would also generate the most output during that time. The average cost of electricity per day as a function of consumption would increase too, since there is a tiered rate structure. Hence, the solar PV system would offset the most expensive electricity and make more economic sense.

Another way of calculating the electricity bill in California is by time of use (TOU). Under this rate structure, the bill will change not only based on season but also on the time of use during the day. The peak rate would apply during the summer months from May to October between 1 P.M. and 7 P.M. In this case, if the electricity usage patterns are skewed toward the later part of the day when the solar PV output decreases, the electricity costs would be much higher.

Economic viability in the United States cannot be generalized even at the city level. A case-by-case analysis is critical before judging the economic viability of the solar PV system. The value of a solar PV system's output is dependent on multiple factors such as the user's electricity usage profile and the electricity rate structure. The actual cost/benefit is dependent on the consumption and the time of consumption of electricity. In California, a solar PV system is economically most beneficial for setups that consume a lot of electricity, such as big homes in hot places like central and southern California.

FINANCING SOLAR PV PROJECTS IN THE UNITED STATES

As PV markets have grown, they have done so through creative financing strategies going from out-of-pocket consumer purchases to more elaborate arrangements. During the home equity credit boom, consumers used their local banks to supply credit as either a HELOC or a second mortgage. In 2006 institutional financiers like Goldman Sachs entered the scene with funding for companies like SunEdison that originated large scale solar projects financed with long-term, fixed-price energy contracts called power purchase agreements (PPAs) signed with utilities, institutions, and corporations. Often these projects financed with PPAs allowed project developers to use the federal tax credits and other subsidy benefits to facilitate several project structures to enable the continued growth of PV in the United States.

Here are some of the recent developments in financing for solar PV projects:

- Third-party ownership model: power from solar PV projects is being sold directly to the end users on a retail basis, through the third-party ownership model.

- Increasingly, the financial sector recognizes solar PV as commercial, reliable technologies.
- Renewable energy certificate revenues are playing a growing role in the success of many solar PV projects.
- Solar REC revenues in states with a solar renewable portfolio standard (RPS) set-aside are particularly important for solar PV development.
- As developers are acquired or team up with larger, better capitalized companies, the financial options available to finance new renewable projects will increase.
- Diversification: investors are diversifying solar PV investment in multiple ways, such as the purchase of structured debt instruments, entering into equity investment partnerships, and possibly partnering with hedge funds.

Renewable energy certificates (RECs) are emerging as a key source of revenues, especially within states that have solar RPS set-asides. Many new market entrants are putting large amounts of capital into the market. A lot of partnerships are also taking place for many new solar PV projects.

Another new emerging business model is the convergence across solar PV and wind markets. Wind developers are entering the solar PV market, and the solar PV developers are adopting financing techniques used by the wind sector.

Third-Party Project Financing: A New Business Model

One of the most promising new business models for solar PV is third-party PV project financing. As opposed to the traditional solar PV model—where the customer purchases a solar PV panel system on his own and has to bear the entire burden—a third party deploys the solar PV system without the customer having to make the full capital investment. In such a model, a big retailer or some other large institution pays for the solar panels for a customer and also signs a PPA to purchase the generated power by that PV system. Often, the price paid for the solar PV generated is slightly lower than the market rate.

More specifically, the solar PV developer installs, operates, and maintains the system on behalf of the customer and the third party. An equity investor buys the project rights from the developer, provides the up-front capital needed to the project's limited liability company (LLC) specifically created for the project, and receives the benefits from the investment tax credits. The project LLC buys the equipment from the manufacturer, and perhaps construction services from the developer. Sometimes the developer might retain ownership of the project until after construction is completed and then sells the project on a turnkey basis. This depends on the cash flow and financial strategic goals of the developer.

So, we see there are four players in this business model:

1. *Host/customer.* The customer does not need to invest any capital. He hosts the solar PV system but does not own it.

2. *Project developer.* The project developer designs, builds, and maintains the solar PV system and also arranges for the transaction and financing.

3. *Manufacturer/installer.* The manufacturer sells and installs the solar PV system and also provides equipment warranties.

4. *Investors.* The investor provides the required capital and owns the equipment. The investor receives the state and federal tax benefits and also receives the income from the electricity sales.

Advantages of Third-Party Business Model

The increasing popularity of this business model can be attributed to the fact that installing solar PV systems costs a lot of money and the third-party model makes it very affordable. Instead of the consumer paying the up-front cost, the third party pays in his place. The consumer continues to pay the regular monthly bill for the electricity to the third party (financier), in place of paying to the utility. The consumer has the advantage of a predictable and fixed-cost supply of electricity, and the third party has the advantage of a steady income stream along with a number of attractive tax credits (in the United States). MMA Renewable Ventures, a San Francisco based company, is one of the leading players in the third-party financing model.

Disadvantages of Third-Party Business Model

One major disadvantage of the third-party financing model is that the solar PV arrangement works best if RECs generated by the solar PV system can be sold to specific buyers of renewable energy or to the utilities that are trying to meet their renewable energy targets. A disadvantage of the third-party business model is that the host company is not able to benefit from the option value of the RECs and in effect sells the environmental benefits to the third party.

Another issue with this business model is if the host goes bankrupt or wants to sell the property with the solar PV system installed. In this case, the investor or financier faces a problem. To tackle it, almost all firms involved in third-party financing have created protections against bankruptcy.

Variations of the Third-Party Business Model

There can be other variations of the third-party business model, such as the lease/buyback arrangement. In this case the host company will also take the responsibility of maintaining and repairing the solar PV panels.

Examples of the Third-Party Business Model

This third-party model is driving significant amounts of capital to the market, a few examples of which include:

- New solar PV projects amounting to $39 million financed by MMA Renewable Ventures in the fourth quarter of 2006. As of the beginning of 2008, its solar portfolio totaled 24.8 MW.
- The $60 million SunE Solar Fund I launched by SunEdison in 2005 to develop 25 projects in the United States, with Goldman Sachs providing the equity and Hudson United Capital (now a unit of TD Bank North) providing construction and term debt financing.
- SunEdison's $26.1 million equity partnership with Goldman Sachs, MissionPoint Capital Partners, and Allco Finance.
- UPC Solar expecting to do big deals by working with owners of large facilities that are willing to host solar PV projects, using at least $50 million worth of solar equipment across multiple properties.

Examples of equity investments in the third-party ownership concept in the past few years include:

- Developing energy-efficient rooftop systems installing approximately one MW of rooftop PV on a General Motors facility in California and expecting to be involved with 50 MW worth of solar roofing projects each year.
- Chevron and Bank of America partnering with the San Jose Unified School District to install five MW of solar PV on the grounds of K-12 schools in California.
- Wal-Mart purchasing power from solar PV projects from SunPower, BP, and SunEdison located on 22 sites, including Wal-Mart stores, Sam's Clubs, and two distribution centers. Total annual production is estimated to be as much as 20 million KWh, which translates into approximately 14.2 MW of total capacity, assuming an average capacity factor of 16 percent.
- Macy's installing 8.9 MW of solar PV on 28 stores across California. In combination with energy efficiency measures executed in these stores, Macy's expects to offset more than 24 million KWh of annual energy consumption. At 11 stores, Macy's owns the solar PV systems outright, while at the remaining 17 stores it purchases electricity generated by SunPower.
- Kohl's signing an agreement with SunEdison under which SunEdison will manage 25 MW of solar PV installed on many of Kohl's stores in exchange for the retailer's commitment to purchase energy from the projects. The installations were expected to be completed in 2008, and the 138,000-plus solar panels are expected to generate more than 35 million KWh annually.

EMERGING BUSINESS MODELS

Deep and volume-driven commodity markets for solar PV in the United States are hindered in part by the lack of significant involvement of utilities to play a role in the development of the industry (that is, systems ownership). Today's solar PV business models

center on user- and third-party-owned solar PV systems. To create large markets for solar PV, the passive role of utilities must
change to accommodate a changing paradigm brought on by the
disruption of their existing business models and to best capitalize
on solar PV's unique benefits. Indeed, at grid-connected rates of
10 to 15 percent of a utility company's peak-load generation, utilities will be driven by concerns for grid infrastructure (a reduction
of traditional capacity additions, reduced substation requirements,
etc.), safety (PV systems generate power back into the grid, thereby
creating sources of hazard in outages for linemen and other utility
workers), and revenue dilution.

Note that much of this section is excerpted from a research
note published by Alternative Energy Investing (AEI).

Role of the Regulator

Solar PV market penetration relies on public policy support, technology, and utility support. As policy, regulatory support and
technology advance utility involvement in the market, the utility
companies represent more than just new sources of large demand.
Utility owned solar PV will unlock additional value not clearly evident today.

By many estimates the cost reduction opportunity in module
distribution, design, installation, ownership, and maintenance
may be substantially reduced by next-generation business models that fully integrate solar PV systems with module supply in a
utility owned model. Such value-enhancing opportunities rely on
more regulatory support for PV, such as uncapped net-metering,
standardized and simplified inexpensive national interconnection standards, and performance-based rate making. These regulatory hurdles remain high, and it is not wise for investors to discount wholesale changes in the next few years. Other regulatory
changes will also take time. Both revenue decoupling to encourage energy efficiency and conservation and tariff structures that
optimize solar PV generation characteristics figure greatly into
changes in regulation to unleash as of yet mostly unrealized value
from solar PV.

Investors must also recognize that policy driven markets require further development of Renewable Portfolio Standards in the United States, especially those with solar PV "carve-outs." At the federal level, greenhouse gas emission reduction programs are required to price emissions more efficiently. Lastly, utilities are waiting for state-level economic development initiatives, and for the growth of state energy initiatives, to provide visibility on returns on investment. Policy changes driving utilities forward are evident; however, we do not expect sufficient momentum or the introduction of new and deep pools of device demand until the middle end of 2010 and beyond.

Utilities Taking the First Step

There is continuing evidence that utilities are expanding their activities and taking the first steps in becoming involved in solar PV. Southern California Edison has been more progressive than most. In a recent announcement, SCE put forth a plan to install solar panels on 150 commercial rooftops over the next five years, creating the world's largest solar PV–generating operation in the world. The project represents 250 MW. SCE asked the California Public Utilities Commission for approval to commit a total of $875 million to the utility's solar project, informing regulators that the expected capacity cost per installed watt would be approximately $3.50, half the average current capacity cost of other photovoltaic installations. Subsequently, on May 8 2008, SCE provided additional cost projections to regulators, forecasting an energy cost of approximately 20 cents per kilowatt hour after adjusting for time of delivery.

SCE intends to invest 4 percent of its five-year capital budget on its solar program. SCE's solar program is likely a first in a move by utility companies to begin changing their business models to own solar PV capacity. SCE has estimated it can earn 12.5 percent ROE, which is 100 b.p. above the CPUC approved rate.

SCE has defined peak capacity as "quick [10 minutes or less] black-start" units and third-party peaking capacity supplied under contract by others. The utility has four 45 MW black-start units (180 MW). However, due to competitive reasons, it is not disclosing its third-party peaking capacity. SCE's peaking requirements in the summer months are fully twice the requirements in the winter months. This is key to understanding the utility's role as a first-mover.

In another important development, on August 14, 2008, Pacific Gas & Electric announced an 800 MW purchase agreement with SunPower (250 MW) and Optisolar (550 MW). Although the headline value of PG&E's announcement was significant for SunPower stock trading over the short term, the announcement had a more subtle value that will take years to realize. The true value to shareholders investing in solar stocks from both SCE's solar rooftop program and PG&E's 800 MW deal is the implied change in business models, where in the next several years utilities will begin to own solar generation, opening large markets for solar PV in the United States and unleash value embedded in solar electric generation currently unrealized in the market. In early 2009, due to the fallout of the credit crisis, Optisolar sold itself to First Solar.

California is home to first-mover utilities for many reasons: a strong RPS, good policy support, and a willingness of the state's PUC to grant cost recovery to solar projects. That said, as noted earlier, both SCE and PG&E do validate a shifting landscape highlighting the U.S. utility market. We think SCE represents a validation of the implied change in utility business models that is most likely necessary to support increased demand and value for solar PV.

Cities Taking Up the Solar Initiative

Cities could take up the initiative to help in the financing of solar PV systems. Berkeley, California, for example, has taken up one such initiative.[3] This plan is referred to as the "Sustainable Energy Financing District" and was granted approval by Berkeley's City Council in September 2008.

The essence of the plan is: Berkeley will take care of up-front financing for PV systems, and the cost will be recovered through property taxes over a period of 20 years. It serves to remove a major hurdle to solar PV because large up-front costs are avoided.

According to the City of Berkeley, an average PV system in the city would cost $28,077. With an average rebate (under the California Solar Initiative) of $6,108, the average cost would therefore be ~$22,000. For a ~$22,000 system, the addition to the property tax every month would be ~$180 ("Berkeley Moves Ahead with Solar-Financing Program," *Business Week*, September 18, 2008). The city would raise cash through bonds to provide for this initiative.

Here are some of the benefits and advantages of this plan as outlined by the mayor of Berkeley:

- It eliminates the need for large up-front costs for solar PV installations.
- The financing obtained by the city would be in the form of a secured bond, with interest rates lower than those available otherwise.
- The repayments on the solar PV system are not tied to the homeowner; rather they are tied to the property, which means the obligations pertaining to the PV system get transferred whenever a new owner moves in.

Such initiatives, though not exactly in the nature of subsidies, can do a great deal to stimulate the solar PV industry.

PV Company Also Doing the Financing

There is the possibility of a new trend where a company has its own financing entity. SunPower signed an agreement with Pacific Gas & Electric through a third-party power agreement counterparty—High Plains Ranch II, SunPower's own financing entity. This enabled SunPower to enter into an arrangement where it will sell only the solar PV electric output and not transfer the ownership of the system itself.

This is a totally different business model from that of a company just selling the modules. This business model is based on selling power from PV modules at a higher rate than the cost of financing the solar PV power plant. It also requires a considerably stronger balance sheet, since the assets (solar PV systems) are now owned by the company rather than transferred to the buyer.

The risk with this business model is that the company would require heavy financing to fund future growth, which would depend on the company's balance sheet to remaining robust.

In this context, we also note some of the other activities SunPower is involved in besides selling and installing solar PV modules.

Financial Consulting

SunPower provides financial consulting for its customers and also helps arrange for finance from companies like GE Commercial Finance and Morgan Stanley. These companies provide services such as project development financing for SunPower's customers. Financial consulting is especially useful because of the large costs to install solar PV systems. By helping customers with this, SunPower can potentially increase its ability to sell solar panels to them.

To detail SunPower's activities further, in the 2007 fiscal year, ~54 percent of the company's systems revenue came from selling solar PV systems to *financing* companies that would in turn enter into power purchase agreements with end customers of electricity. These power purchase agreements (PPA) could be based on savings on electricity consumption as a result of installing solar PV panels. In fact, fully 16 percent of SunPower's systems segment revenue for FY2007 came from a single financing company: MMA Renewable Ventures. A potential downside to this kind of business model is that if financing companies face a difficult credit environment, sales through this route could slow down considerably.

Maintenance/Monitoring

SunPower offers two levels of service and monitoring for its customers, namely, standard service and plus service. SunPower also has a proprietary data acquisition system that allows it to monitor and check the performance of installed systems. And it has a Web site, "SunPowerMonitor.com," where customers can access data on how the solar panels are performing.

In addition, SunPower offers several other services, including consulting on energy efficiency. Company projects include:

- Upgrade of heating, ventilation, and air-conditioning systems
- Reducing electricity use by using variable frequency drives
- Energy-efficient lighting
- Developing energy management systems
- Energy management and diagnostics in buildings

These indicate that in the future some solar PV companies like SunPower may seek to become service providers at a broader level, rather than just selling modules.

Solar Business Model Case Study: Selco India

Selco,[4] of Bangalore, India, markets solar PV systems to low-income households and also to institutions that are not connected to the grid. With an annual turnover of ~£1.5 million, Selco is certainly not a large company in the context of the solar PV industry. However, its choices of business model and target market are unique and therefore worth examining in some detail.

To give a bit of background on this, large parts or rural India are not connected to grid electricity. In this context, Selco markets so-called "solar home systems" (SHS), consisting of a 35 W_p roof-mounted module plus a lead-acid battery to store electricity (this battery typically lasts ~five years). These systems are primarily used to provide lighting, but also at times to power radios and fans. Selco also takes care of customizing the installation according to the needs of the customer.

A typical (four-light) solar home system from Selco costs ~INR 18,000 (~$400, assuming an exchange rate of 45 INR to the dollar), and requires a small down payment with monthly installments of INR 300 to 400 (~seven to nine dollars) over approximately five years. If the lighting provided by Selco's solar PV systems allows households to work extra hours (such as on tailoring or basket-making) and generate enough income to offset the monthly installments for the SHS, they would find it economical to make the purchase.

While Selco does not directly offer financing for its customers, it does have relationships with finance institutions so it can help customers gain access to credit. Also, it provides "down payment guarantees" to help with financing in cases where customers are unable to provide the required up-front payments (these usually amount to ~15 percent of system costs). For its innovations in bringing solar PV to rural areas without electricity access, Selco was awarded the Ashden Award first prize in 2005 for sustainable energy.

Speaking more generally, Selco's example demonstrates the possibility of solar PV emerging as a potential source of electricity in remote areas that otherwise have no access to electricity. Solar PV based electricity can be generated and consumed on-site and does not need costly supporting infrastructure: these characteristics make solar PV particularly attractive in such places. As system prices fall further, the opportunities to serve this market would increase substantially.

N O T E S

1. Suntech's diversified polysilicon sourcing strategy included an investment in Hoku. Company executives characterized the investment as "an option" having high return, high risk, and modest payoff potential.
2. http://files.shareholder.com/downloads/FSLR/0x0xS950157-07-864/1274494/filing.pdf
3. http://www.ci.berkeley.ca.us/mayor/GHG/SEFD-summary.htm
4. http://www.ashdenawards.org/files/reports/SELCO2007technicalreport.pdf

CHAPTER 4

Solar Stock Valuation

Valuation is an art, not a science. There is no right or wrong measure for valuing solar stocks. That said, traditional valuation measures such as price-to-sales and price-to-earnings, used in relative valuation, are often used to decide which solar stocks seem attractive and which do not.

The opportunities revealed by valuation are only half the story, however. To fully unlock the profit-making potential of solar stocks, investors must understand how underlying fundamentals drive the estimating process and why sales and earnings of the solar industry are impacted by forces not commonly found in other industries. While the valuation principals do not change from industry to industry, the estimating process used to predict future performance of solar stocks does. In this chapter we will first demystify valuation and then illuminate the forces that drive the solar business.

RATIO-BASED ANALYSIS

One way to value solar stocks is to compare their ratios, such as price-to-earnings (P/E), price-to-sales (P/S), price-to-book (P/B), EV/EBITDA (enterprise value divided by earnings before interest, taxes, depreciation, and amortization), return on capital employed (ROCE), and return on equity (ROE).

One advantage with these metrics is that though relatively simple to calculate, they give an investor some degree of insight into whether stocks are fairly valued. In describing these metrics, we first start out with the principles behind them and the various ways in which they can be applied to solar stocks. Valuation ratios can be compared to peer companies and also to broader market indices such as the Nasdaq Composite, S&P 500, and Russell 2000. The questions an investor would typically ask are:

1. Are the valuations comparable to those of the peer group?
2. If the valuation of a company is significantly higher than the average for its peer group, do the company's fundamentals and growth prospects justify the valuation? Here an investor in a solar company should ask: Is there something exceptional about the company that makes it stand out from its peers? Does it have a technology or manufacturing-related advantage? Is the high valuation related to advantageous sales contracts or raw material contracts? More important, the investor would want to find out if this advantage is sustainable or likely to disappear over the long run.
3. If the valuation of a company is significantly lower than peer valuations, the investor must ask whether the company's prospects are unfavorable enough to justify the lower valuations. Are the low valuations related to strategic moves (such as vertical integration into polysilicon) that have not convinced investors? Are they because of doubts related to technology (such as upgraded metallurgical grade silicon)? Are problems with the company of a long-lasting nature or likely to diminish over a slightly longer time frame? If the investor concludes that there is no such fundamental reason justifying the low stock price, this may be a buying opportunity.

As with any relative valuation exercise, an investor needs to be prudent in applying this methodology because peer valuations may themselves not truly reflect fundamentals. For instance, if the peer group is itself overvalued, using this as a benchmark to estimate the value of a solar stock would not be correct. It is useful, therefore, to also compare peer group valuations with broader market indices, such as the Nasdaq Composite and S&P 500, to estimate if these valuations are in line.

At times these broader market indices might also be overvalued or undervalued as compared to fundamentals. For example, if the equity markets are overheated for a while, the P/E or P/S of broader market indices would not truly reflect fundamentals. Therefore, as a further check, an investor may want to compare relative valuations with historical ratios for these indices; this helps keep relative valuation estimates anchored as much as possible to fundamentals. Again, valuation is as much an art as a science, and such checks help an investor gain a better perspective on where the valuation of a solar stock stands.

Let's take a closer look at some commonly used ratio-based valuation metrics and the various ways in which they can be applied to solar stocks.

Price-to-Earnings Ratio

The price-to-earnings (P/E) ratio is calculated as the stock price divided by the trailing-12-months earnings per share (EPS). This ratio could be applied as a principle for investing at different levels:

The average P/E of the stocks in the solar sector can be compared with the P/E of stock indices such as the S&P 500 and the Dow Jones Industrial Average. This would give the investor an idea of how expensive or cheap the solar sector is relative to the market. The stocks in the solar sector can be grouped according to technology and positioned in the supply chain. Now we can compare the P/E ratios of each group to help us decide which are more attractive relative to others.

One issue with the price-to-earnings ratio for solar stocks is the fact that some of these companies could be at a nascent stage where they have not yet shown profits. For such companies, the price-to-sales ratio might be a better indicator of how attractive the stock is.

Price-to-Sales Ratio

The price-to-sales (P/S) ratio for a company is calculated as the market capitalization of the company divided by trailing-12-months revenue. This ratio has the advantage that it can be computed even for companies with negative earnings. Further, from an investor's perspective, while "earnings" numbers often depend on specific accounting policies adopted by the company, "revenue" numbers are usually more consistent when comparing different companies.

An investor looking at the solar sector could apply the price-to-sales ratio similar to the way the P/E ratio is applied, as follows:

1. The average P/S of the stocks in the solar stock universe can be compared with the P/S of stock indices such as the S&P 500 and the Dow Jones Industrial Average.
2. The stocks in the solar sector can be grouped according to technology and positioned in the supply chain similar to the way "price-to-earnings ratio" principle is described. Now we can compare the P/S ratios of each group.
3. When adopting a portfolio approach to investing in the sector, we could use P/S ratios to decide which specific stocks to pick in each group.

EV/EBITDA

To understand this ratio, you need to first understand how to calculate enterprise value (EV). Here is how Investopedia.com explains EV:

> [Enterprise value is] a measure of a company's value, often used as an alternative to straightforward market capitalization. EV is calculated as market capitalization plus debt, minority interest and preferred shares, minus total cash and cash equivalents.

EV/EBITDA, as noted earlier, is the enterprise value divided by earnings before interest, taxes, depreciation, and amortization. This ratio is also referred to as the "enterprise multiple." In essence, the enterprise multiple is a measure of how much a company is worth. This measure is often used to calculate how much it would cost to acquire a company when that company's debt is taken into account.

The advantage of EV/EBITDA is that it allows comparison of companies with different capital structures. Similar to the P/E or P/S, we can use the peer group EV/EBITDA to evaluate the EV/EBITDA for any solar company. The enterprise multiple can also be used with forward estimates of EBITDA from Street consensus (although in this case, there is a risk of the consensus estimates being incorrect).

From the EV estimated using peer comps, we can obtain an estimate of the target price. For solar companies, a simple way to arrive at comparable companies is to group them according to their position in the supply chain, as we do below (EV to trailing 12-month EBITDA from Bloomberg, as of September 19, 2008).

Polysilicon/Wafer Companies

- MEMC: 12.66
- Wacker: 6.58
- Sumco: 3.91
- DC Chemical: 16.03

Cell/Module Companies

- Q-Cells AG: 26.67
- First Solar: 67.43

Q-Cells' lower EV/EBITDA compared to First Solar may be due in part to Q-Cells' dominant c-Si core business versus First Solar's pure-play cadmium telluride technology, which boasts significantly lower cost and a relatively short time to achieving grid parity.

System Integrators/Installers

- Phoenix Solar: 9.94

Vertically Integrated

- Renewable Energy Corporation: 20.98
- Solarworld AG: 11.25

Renewable Energy Corporation's vast polysilicon manufacturing base, enhanced by multiple low cost FBR technology, combined with integrated ingot, wafer, cell, and module operations, is valued more highly than Solarworld, which only recently engaged in a joint venture with Evonik to develop a polysilicon facility. The new JSSI joint venture polysilicon technology between Solarworld and Evonik may be more energy efficient and less costly compared to Renewable Energy Corporation's FBR and Siemens' technologies. Should the JSSI technology bear fruit, Solarworld could see multiple expansions rewarding investors with a higher stock price.

Solar Capital Equipment

EV/EBITDA can be compared with ratios in other industries as well (wind energy, semiconductor industry, etc.) in order to see whether valuations mirror fundamentals. We can also compare this with the ratios for the Dow Jones Industrial Average, S&P 500, and other key benchmark indices as an additional check on the company's valuation.

- Amtech Systems: 15.29
- Centrotherm: 46.51

Centrotherm has an installed base of c-Si solar cell customers throughout the world, and its size in sales and earnings dwarfs that of Amtech Systems. Centrotherm is a vastly superior company, with emerging sales coming from a new turnkey offering that utilizes thin film CIGS technology.

The disadvantage of using EV/EBITDA is that its calculations are more involved than a simple ratio such as P/E or P/S.

Price-to-Book Ratios

A company's price-to-book (P/B) ratio is calculated by dividing its market capitalization by the company's total book value from its balance sheet. Below we list the P/B values of some solar companies by their position in the supply chain (data as of August 2008).

Polysilicon/Wafer Companies

- MEMC: 5.26
- Wacker: 3.32
- Sumco: 1.56
- DC Chemical: 5.68

Cell/Module Companies

- First Solar: 17.38
- Suntech Power: 6.1
- JA Solar: 4.13
- Evergreen Solar: 1.92

System Integration/Installation

- Akeena Solar: 4.16
- Phoenix Solar: 6.29

Vertically Integrated

- Yingli Green: 3.34
- Trina Solar: 1.99
- REC: 6.53
- Solarworld AG: 5.05

Solar Capital Equipment

- Spire Corporation: 10.62
- Amtech Systems: 1.31
- OC Oerlikon: 2.75

VALUATION WITH M&A DEALS AS REFERENCE

We can use recent mergers and acquisitions (M&A) deals in the solar industry as reference points on valuation. These are very useful because they give actual multiples that were paid out for acquiring stakes or taking over companies. It is important here to ensure that the multiples from these M&A deals are applied to the *right* kind of companies. For example, if the takeover target is a vertically integrated company, the valuation multiples obtained as a result of the deal would be suitable for a similar, vertically integrated company but not for, say, a systems integration company.

Robert Bosch's acquisition of ersol Solar Energy AG in June 2008 can serve as a guide on valuation for companies in the sector. The first step is to identify the business activities ersol is involved in, so we only use this as a reference point for similar companies.

Ersol is a fully integrated silicon-based solar company (not silicon production, but silicon recycling), producing wafers, cells, and modules. Knowing this, we can readily identify SunPower and Evergreen Solar as U.S.-traded solar companies with a scope of business activities similar to that of ersol Solar Energy.

Bosch acquired ersol Solar Energy for $1.67 billion, a figure 6.8 times more than the FY07) revenue, 48.5 times FY07 EBIT, and 124 times FY07 earnings. For SunPower, applying these multiples to FY07 revenue, EBIT, and earnings gives an implied floor value of $5.27 billion, $1.26 billion, and $1.14 billion, respectively, as depicted in Table 4.1.

We can also obtain implied floor values on a "per share" basis. The "implied floor prices" on the basis of FY07 revenues, EBIT, and net profit are $131.3, $31.4, and $28.4, respectively. Similarly, when we apply the revenue multiple (6.8 times) to Evergreen Solar's FY07 revenue, we find that the implied floor value is $475.1 million and the implied floor price is $3.1/share, as can be seen in Table 4.2.

T A B L E 4.1

Estimates of Implied Floor Value of SunPower

2007	$mn	Multiple	Floor Value ($mn)	Wtd. Avg. Shares Outstanding, Basic (mn)	Implied Floor Price/Share ($)
Revenues	774.79	6.8	5,268.6	40.1	131.3
EBIT	25.98	48.5	1,260.0	40.1	31.4
Earnings	9.20	124	1,140.8	40.1	28.4

Source: SunPower 10-K filings, Bloomberg

T A B L E 4.2

Evergreen Solar

2007	$mn	Multiple	Floor Value ($mn)	Wtd. Avg. Shares Outstanding, Basic (mn)	Implied Floor Price/Share ($)
Revenues	69.87	6.8	475.1	152.2	3.1
EBIT	−25.58	48.5	n/a	152.2	n/a
Earnings	−16.60	124	n/a	152.2	n/a

Source: Evergreen Solar 10-K filings, Bloomberg

In the case of Evergreen Solar, we cannot apply the multiples on EBIT and earnings because these numbers are negative for FY07.

In comparing ersol Solar Energy, SunPower, and Evergreen Solar, we see that the June 2008 acquisition was in all likelihood transacted at too high a price, given that by year end the applicable peer group was trading far below comparable multiples when the BOCH acquisition of ersol Solar Energy was priced.

DISCOUNTED CASH FLOW VALUATION

Discounted cash flow (DCF) valuation is perhaps the most important way to value solar stocks. We first explain the essence of our DCF valuation methodology and then present an example using an imaginary company ("Company Alpha") to illustrate the calculations and the particular questions solar investors must ask themselves as they estimate the inputs.

As we shall see, the DCF valuation methodology is much more involved than other approaches to valuation, but it has the advantage of being based completely on projections of a company's fundamentals. When the stock price is below what a well-conceived DCF would suggest, investors would be expected to buy. Conversely, when the share price is above the DCF value, investors would be expected to sell. On the other hand, relative valuation depends on peer companies being fairly valued on average. There may be situations when the industry P/E or P/S ratios are totally out of sync with fundamentals; indeed, before the famous dot-com crash that started in early 2000, relative valuation approaches would have yielded highly inflated target prices for most Internet stocks.

Our DCF valuation involves finding the net operating profit after tax, or NOPAT, which is calculated as EBIT times (1 – Tax rate) for each year. "Free cash flow" for each year is then calculated as NOPAT minus the increase in invested capital.

To account for the time value of money and risks, we discount the free cash flows by the weighted-average cost of capital (WACC) for the company. The WACC indicates the required return on the invested capital, given the company's costs of equity and debt, as well as its financing mix.

WACC is calculated as $(E/V) \times R_e + (D/V) \times (1 - T_c) \times R_d$, where R_e and R_d are the costs of equity and debt, respectively, and E/V and D/V refer to the proportions of equity and debt, respectively, in the financing mix. T_c stands for the marginal corporate tax rate, and the $(1 - T_c)$ accounts for the tax-saving effect of interest expenses on debt. The cost of equity, R_e, is calculated by the capital asset pricing model (CAPM), using the formula $R_e = R_f + \beta (R_m - R_f)$, where R_f stands for the risk-free rate, β is the beta of the stock, and R_m is the required return on the market portfolio (that is, the required rate of return on a portfolio meant to represent the entire market).

Solar investors must consider the global nature of a company's business model and determine the degree to which the company's revenues are concentrated in one geography or another when calculating the cost of equity. Since most solar companies have sales and revenues in Germany, the world's largest solar market, the riskless rate should consider blending the German risk-free rate into the calculus. Betas, too, need to be considered in the context of the solar industry. For example, the solar industry is a subsidized industry that will remain dependent upon these subsidies until grid parity is achieved. If no subsidies exist, the industry suffers a loss of what is potentially most of its demand.

Second, operating leverage in the solar industry can also be difficult to estimate, due again to the subsidy dependent nature of the industry. For example, cyclical forces in a subsidized market can be exacerbated since companies seeking to enter the market or looking to expand tend to face greater periods of oversupply as they engage in excessive build-outs. This is particularly the case when subsidies are too high or are without limitation. Solar companies have increased their operating leverage with investments in costly equipment to advance scale. As a result, they have bet on continued growth in sales and earnings, leading to more investments and variation in volatility.

Third, because most solar companies in the public markets are new, there is little statistical data for estimating beta and other parameters relative to the cost of capital. Since this is the case, it is difficult to compare companies to each other with any statistical significance. Investors must therefore strongly consider determining the beta variables for the cost of equity calculation using fundamental factors.

Upon calculating the cost of debt, another essential factor input into the cost of capital, investors must consider carefully if the default spread over the risk-free rate reflects the company's risk. For solar companies, the estimator must judge a reasonable default spread encompassing several stages of economic growth. The default spread in periods of shortages of solar materials and devices is usually accompanied by higher selling prices and industry profits somewhere along the supply chain. Thus, market rates for debt are likely to be lower than on average, when polysilicon and modules are in ample supply, prices are down, and industry profits are down, too. As is the case in 2009, the industry is amply supplied with everything from polysilicon to modules; therefore, profits are down greatly, and default spreads have blown out.

Our WACC calculations for Company Alpha are presented in Table 4.3.

In our example, the DCF valuation is done in three stages. We will first explain these stages and then describe our DCF calculations.

Stage 1

The first stage is based on explicit forecasts of all financials for the next three years, plus the forecasts for the remainder of the current year. For example, if our valuation is being done in August 2008, the explicit forecast period would include the years 2008, 2009, 2010, and 2011. This is because we believe it is possible to reasonably estimate financials over this relatively short time period.

For PV manufacturing companies, revenue estimates would be based on projections of megawatt shipments and average selling

T A B L E 4.3

WACC Assumptions for Company Alpha

Risk-free rate (R_f)	4%
Equity beta (β_e)	1.6
Equity risk premium ($R_m - R_f$)	5.0%
R_e (= R_f + beta × ($R_m - R_f$))	12.0%
R_d (cost of debt)	5.0%
Target D/V	40.0%
Target E/V	60.0%
WACC	8.7%

TABLE 4.4

Free Cash Flow Projections for Stage 1

	Reference 2007	2008	2009	2010	2011
Revenue	$100.0	$120.0	$140.0	$160.0	$180.0
% change in revenue yoy		20%	17%	14%	13%
Cost of revenues	$75.0	$92.0	$103.0	$120.0	$136.0
Gross margin	25.0%	23.3%	26.4%	25.0%	24.4%
Operating expenses	$15.0	$18.0	$21.0	$24.0	$27.0
As % of revenue	15.0%	15.0%	15.0%	15.0%	15.0%
EBIT	$10.0	$10.0	$16.0	$16.0	$17.0
Cash tax rate	25%	25%	25%	25%	25%
NOPAT	$7.5	$7.5	$12.0	$12.0	$12.8
Invested capital	$150.0	$160.0	$170.0	$180.0	$190.0
% increase in invested capital		6.7%	6.3%	5.9%	5.6%
Increase in invested capital		$10.0	$10.0	$10.0	$10.0
Free cash flow		–$2.5	$2.0	$2.0	$2.8

prices (ASPs). Cost of sales can then be estimated using projections of gross margins. Similarly, all relevant line items are estimated.

Table 4.4 presents our estimates and calculations for Stage 1, with all numbers representing millions of dollars, except for those indicating percentages and per share prices.

Stage 2

Projections for the next five years are based on estimates of percentage growth in revenue, gross margins, tax rates, and growth in invested capital, rather than explicit forecasts for each line item. Apart from this, the calculations are very similar to those in Stage 1. Table 4.5 illustrates Stage 2 of the DCF valuation, with numbers representing millions of dollars except for percentages and per share prices.

Stage 3

Beyond the year 2017 we assume that free cash flows keep growing at a constant rate of 3 percent until perpetuity. This assumption is used to derive the "terminal value" of the cash flows.

TABLE 4.5

Free Cash Flow Projections for Stage 2

	2012	2013	2014	2015	2016
Revenue	$198.0	$217.8	$239.6	$263.5	$289.9
% change in revenue yoy	10%	10%	10%	10%	10%
Cost of revenues	$149.6	$164.6	$181.0	$199.1	$219.0
Gross margin	24.4%	24.4%	24.4%	24.4%	24.4%
Operating expenses	$29.7	$32.7	$35.9	$39.5	$43.5
As % of revenue	15.0%	15.0%	15.0%	15.0%	15.0%
EBIT	$18.7	$20.6	$22.6	$24.9	$27.4
Cash tax rate	25%	25%	25%	25%	25%
NOPAT	$14.0	$15.4	$17.0	$18.7	$20.5
Invested capital	$199.5	$209.5	$219.9	$230.9	$242.5
% increase in invested capital	5.0%	5.0%	5.0%	5.0%	5.0%
Increase in invested capital	$9.5	$10.0	$10.5	$11.0	$11.5
Free cash flow	$4.5	$5.5	$6.5	$7.7	$9.0

We add up the discounted cash flows from Stages 1, 2, and 3 to obtain the present value of future cash flows. To this we add "cash and equivalents" and subtract the value of debt to get the value of equity. The value of equity divided by the number of shares gives the target price per share, as shown in Table 4.6.

TABLE 4.6

Calculation of Target Price for Company Alpha

DCF for Stage 1	$2.9
DCF for Stage 2	$18.1
DCF for Stage 3	$76.6
Firm value	$97.6
Add cash	$100.0
Less debt	$50.0
Equity value	$147.6
Number of shares outstanding (million)	10
Per share value	$14.8

T A B L E 4.7

Target Price Sensitivity to Different Values of WACC and the Terminal Growth Rate (g)

g \ WACC	7.0%	8.0%	9.0%	10.0%	11.0%	12.0%
2.0%	$17.3	$14.8	$13.1	$11.8	$10.8	$10.0
3.0%	$19.9	$16.5	$14.2	$12.5	$11.3	$10.4
4.0%	$24.3	$18.9	$15.7	$13.5	$12.0	$10.9
5.0%	$33.0	$22.9	$17.9	$14.9	$13.0	$11.6

From our DCF calculations we arrive at a target price for Company Alpha of $14.8/share.

Table 4.7 highlights the sensitivity of the target price to different values of WACC and the terminal growth rate (g):

We see from this table that the target price is in the range $10.0 (when WACC = 12%, g = 2%) to $33.0 (when WACC = 7%, g = 5%), for the values of terminal growth rate and WACC that we have considered.

Potential Issues in DCF Valuation for Solar Companies

While a discounted cash-flow-based valuation for a company is useful and rooted in the fundamentals of the company, performing the necessary computations represents a particular challenge in the case of the fast-growing solar industry. Indeed, it is very difficult to make reliable predictions of revenues, earnings, and cash flows more than a few quarters into the future. Moreover, there are several companies that have a history of losses or have only recently started making profits, with amorphous-silicon-based Energy Conversion Devices a case in point. In such cases it is even more difficult to predict future earnings apart from the guidance given by management. Estimation of the "terminal growth rate" (used in Stage 3 of the valuation methodology described in this chapter) is also not an easy task.

However, these shortcomings can be mitigated to some extent by a sensitivity analysis similar to the one we have outlined. By analyzing "worst case" and "best case" scenarios using this analysis, an investor would be able to make a more nuanced call on the stock.

DCF valuation can also be used as one of the inputs in a "blended" valuation, where relative valuation-based metrics can serve as other inputs in the process. For example, an investor may want to capture the information available in a DCF valuation together with the information in a P/S- and P/B-based valuation. The investor may do so by calculating the value of the stock as a 40-30-30 percent blend of DCF, P/S, and P/B, respectively. Such a "blended" approach to valuation ensures that information from multiple methodologies is captured, and it also mitigates the possibility of errors in the valuation number.

An investor might also find it useful to just compare the DCF-based "target" stock price with the target prices obtained by relative valuation to check for any anomalies.

EARNINGS DRIVERS

In this section, we describe the key earnings drivers for companies in the PV industry. As we identify various earnings drivers, we also present examples of solar PV companies with regard to them.

Average Selling Prices

Estimating selling prices, half of the revenue calculation, is largely dependent on supply and demand in almost all industries. It is no different in solar. However, in the solar industry the reliance on government subsidies could accelerate changes to selling prices. This appears to be the case because the subsidy acts on demand in relation to certain price floors and ceilings established as a basis of project investors' hurdle rates. This dynamic is itself a source of volatility that we will elaborate on further, in the section on margins.

Some solar companies can charge a premium on PV products due to their differentiating characteristics. For example, SunPower has higher average selling prices (ASPs) than competitors due to (1) higher efficiency, which allows more power output for the same installed area and (2) the superior aesthetics of its panels. On the other hand, Energy Conversion Devices charges lower ASPs as a thin film company despite its unique flexible, lightweight (just 1 lb/sq ft, as compared to ~2.2 to 5.0 lb/sq ft for a glass substrate) amorphous-silicon-on-stainless-steel panels, that specially address the BIPV market.

Since the market for PV is still largely driven by government subsidies, any changes in these could significantly affect ASPs. For example, higher degression rates on feed-in tariffs would bring down ASPs. A company's geographic mix of revenues is therefore a major determinant of ASPs. First Solar, for instance, has over 80 percent of its revenue coming from Germany, so ASPs (and therefore earnings) would be heavily influenced by the feed-in tariff regime in Germany.

Type of customer/end-market mix could also determine ASPs. For example, companies such as Energy Conversion Devices, with a large portion of its revenues coming from BIPV, would benefit from higher feed-in tariffs and other forms of subsidies for BIPV.

Gross Margins

In the section above on selling prices we introduced the idea that estimating sales and margins in the solar industry is closely tied to subsidy support, at least until the industry matures and grid parity is achieved. The effect on margins from these dynamics is a topic for any estimator worth his salt to consider in great detail. Until recently some estimators have relied on simplifying the process by clinging to the statement that because Germany is an unlimited subsidy program, the industry cannot be oversupplied. We think the closer an estimator looks at this, the more questions will be raised, leading the more thorough estimators to resist the notion that because of Germany's uncapped market, there cannot be long periods of surplus materials and devices.

Comparing Pricing Power and Cost Control

In the absence of other factors, as described above, we will now take a look at comparing average selling prices and variable costs among companies that employ a specific technology so we can get an idea of how each company fares on pricing power and cost control. Gross margin is a parameter that captures these.

This comparison needs to be done at the most detailed level possible. That is to say, we need to group companies according to technology so every group contains only companies with closely similar technologies and is at the same position in the supply chain. The idea here is that we might still want to invest in a company with relatively low current gross margins if we believe that it employs significantly different technology that could emerge a winner in the future.

For example, rather than compare the gross margins of all thin film companies, it would be better to group companies separately on subtechnology, like cadmium telluride (CdTe) and amorphous silicon (a-Si). Similarly for companies based on crystalline silicon, we would want to consider the monocrystalline and polycrystalline players separately.

To illustrate, in an analysis by Alternative Energy Investing, gross margins for companies making modules based on monocrystalline silicon averaged 19 percent, polycrystalline ~19 percent, amorphous silicon thin film ~31 percent, while wafer manufacturers had an average gross margin of 25 percent. In each of these groups, we would want to select the companies with the best gross margins.

If we look at gross margins from a supply chain perspective, meanwhile, we see that silicon companies have enjoyed a huge boost in margins since 2004, thanks in large part to increased demand, as the following chart in Figure 4.1 illustrates. We may also see, as in Figure 4.1 that those industry consultants who endorse the notion that uncapped subsidies foster increasing margins over time without protracted periods of oversupply are wrong. The investor must beware of consulting companies that align themselves with

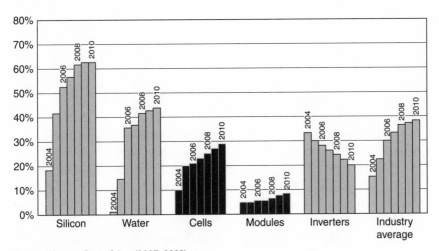

Source: Photon Consulting (2007–2008)

Figure 4.1 Pretax profit margin breakdown, 2004–2010 estimated

media outlets to ensure that the "hype" in solar does not overwhelm their own judgment and common sense.

However, with likely silicon oversupply, it's quite possible that its margins will shrink. Meanwhile, Figure 4.1 shows that overall the solar PV industry has enjoyed a healthy growth in margins from 2004 to 2007, with some estimators predicting this will continue through at least 2010. With the credit crisis in full swing during the publication of this book and the oversupply of practically everything in the solar supply chain that began in the second half of 2008, company margins have collapsed taking with them the notion that a dramatic oversupply would not occur in the presence of "backstop markets" (i.e., Germany with its generous and uncapped subsidy program).

Take-or-pay supply and sales contracts struck at boom-time high prices and a "backstop market" demand driven ideology are today pulling down margins and may ultimately be recorded as the critical failure of the industry to stick to supply chain basics. Companies that can renegotiate these uneconomical contracts, or break them outright, may return to a normal margin structure (i.e., 10 percent to 15 percent for wafer companies, 15 percent to 20 percent for cells, and 5 percent to 7 percent for modules) and their stocks may benefit most as the oversupply abates in the years ahead.

Costs

Given the tight supply of polysilicon in recent years, costs for crystalline-silicon-based companies are determined to a large extent by how companies manage their raw material costs. The position of a solar company with regard to raw material supplies is therefore a major driver of earnings until poly supply eases.

In recent years companies that have been well-positioned in terms of their raw material contracts have had a major advantage in costs, compared to companies that depend on the spot market. As we shall describe in Chapter 6, several companies have long-term contractual agreements with raw material suppliers that help them manage these supplies at reasonable prices.

In the solar PV industry, where raw material costs are such an important portion of the cost of goods sold (COGS), costs (and therefore earnings) can be heavily impacted by the nature of a company's raw material supply arrangements. Different companies tackle this issue in different ways. Suntech Power has signed several long-term supply contracts to take care of its raw material needs. Similarly, Evergreen Solar has also signed multiyear agreements with several companies, such as Wacker and DC Chemical, to secure its raw material supplies. SunPower also has a long-term contract with Hemlock to obtain silicon raw material.

On the other hand, some companies may choose to go in for vertical integration to secure their raw materials. Of course, this strategy comes with several execution-related risks since polysilicon and wafer manufacturing involve relatively complex manufacturing processes. Besides these, companies could also cut down on silicon costs by using more material-efficient technology and manufacturing processes. For example, Evergreen Solar uses its proprietary string ribbon technology, which reduces silicon consumption and therefore lowers silicon costs per watt. This reduces the company's silicon consumption to < 5g per watt.

With the increasing adoption of the upgraded metallurgical silicon route for PV manufacturing, companies could potentially lower raw material costs. For example, Q-Cells has signed an agreement with Canada-based Timminco for procurement of upgraded metallurgical silicon and plans to have ~50 percent of its silicon supply from upgraded metallurgical silicon by 2010. If upgraded metallurgical silicon-based cell manufacturing takes off without major problems, this could help decrease costs until more poly supply comes online.

For thin films, cost advantages could be achieved (among other methods) through manufacturing innovations such as roll-to-roll manufacturing. Energy Conversion Devices, for instance, has a highly automated, proprietary roll-to-roll thin film process that is geared toward low cost manufacturing. Indeed, the company has a road map for cost reduction to achieve grid parity by 2012. One-third of the cost reductions to achieve this would come from material sourcing, one-third from manufacturing efficiencies, and one-third from improvements in areas such as conversion efficiency.

Capital costs per watt are important because they give an indication of how efficiently the company can scale up its operations in the future. To apply this principle, we first need to group companies in detail by technology, similar to the classification for the gross margin principle. The idea here is to be able to compare fixed costs per watt within each technology group. We could also compare capital costs per watt across technologies.

Capacity Expansions

Given the fast-growing demand for solar PV, the ability to ramp up capacity quickly and in large increments assumes critical importance. Indeed, capacity addition is a key driver of revenue and earnings. The ability of a solar company to ramp up is a function of several factors.

In terms of financing capacity expansions, capital costs per watt give an indication of how easy or difficult it is to support capacity expansions. Suntech Power has a capex-per-watt figure of just $0.27/W (cost of land, buildings, equipment on a "cells + modules" basis). This means that if Suntech Power decided to expand capacity by 500 MW, it would incur a cost of ~$135 million. On the other hand, Evergreen Solar's capital costs per watt are relatively high at ~$2.33/W (of which ~61 percent, or $1.41/W, is attributable to equipment costs) partly because it requires specialized equipment for its string ribbon manufacturing process. We note, however, that capital costs for Evergreen Solar are expected to come down to around $1.22/W by 2010, out of which $0.83/W would be for equipment and $0.39/W for buildings.

In this context, we observe that if a PV company's manufacturing process requires customized, nonstandard equipment, capital costs per watt would usually tend to increase as compared to using off-the-shelf equipment.

If a company has a clear method to replicate production lines, it can prove to be a significant advantage in expansions. For example, First Solar employs a "Copy Smart" methodology that makes it easier to add new production lines and therefore helps accelerate capacity additions.

Availability of capital is another determinant of how easily capacity can be added. If a company is already significantly leveraged, it may be difficult to raise further capital at reasonable costs to fund expansions. Financial leverage can be measured by the debt-to-equity ratio. A lower debt-to-equity ratio would indicate a more conservative profile for an investor, in the sense that such a company is less likely to go bankrupt. However, for a given return on assets, a company with a lower debt-to-equity ratio would have a lower return on equity. An investor would need to decide on the trade-offs in this regard.

Given the current situation in the solar industry, where a company inherently faces technology risk, an investor might want to avoid further risk on the financial side by selecting companies with a relatively low debt-to-equity ratio. In doing so, the investor would be implicitly relying on the company's operational expertise and ability to employ capital efficiently, to generate good returns, rather than on financial leverage.

The following debt-to-equity ratios, defined here as "short-term + long-term borrowings," are given as a percentage of "total common equity":

- Sunpower (SPWR): 49 percent
- Evergreen Solar (ESLR): 23 percent
- First Solar (FSLR): 10 percent
- Energy Conversion Devices (ENER): 5 percent
- JA Solar (JASO): 5 percent

A company's record in terms of capacity ramp-ups gives an investor insight into its earnings power. In the case of Energy Conversion Devices, ramp time has decreased from 37 months (for its Auburn Hills 1 facility) to just nine months (for its Greenville 1a facility).

JA Solar is another company that has managed to add capacity rapidly. As of April 2006 the company had just a 25 MW line; within a matter of a few months (by October 2006), JA Solar had increased capacity to 175 MW. This rapid ramp-up is set to continue, with 425 MW of capacity expected by the end of 2008.

ROIC/ROER VALUATION

As mentioned at the beginning of this chapter, valuation is as much an art as it is a science. An investor can, however, use several measures to determine the fundamentals of the company. In particular, an investor would want to know how efficiently the company is using capital and the returns generated on equity.

Return on invested capital (ROIC) and return on equity (ROE) measures can be used to supplement valuation metrics in making investment decisions vis-à-vis solar PV companies. Another point to note here is that these return metrics provide a tool for the investor to gauge which parts of the solar chain are generating more value as compared to others. As we explain in detail in Chapter 8, if an investor is adopting a top-down portfolio management strategy with allocations based on position in the supply chain, these metrics can be used in conjunction with knowledge of the industry fundamentals to estimate where maximum value is being created and to modify allocation accordingly.

Specifically, given the tightness in polysilicon in recent times, much of the value in the supply chain has been captured in the last few years by polysilicon companies. As polysilicon supplies ease and new markets for solar PV emerge, this situation might change, and the investor would have to take this into consideration.

Besides valuation and ROIC/ROE numbers, an investor would be interested in knowing what other characteristics distinguish one company from another in different parts of the supply chain. For example, what parameters should an investor use to judge whether one company in the polysilicon portion of the supply chain is better than another polysilicon company? Or, when does she decide that one solar PV module company is better than another?

Below, grouped by their position in the supply chain, are the returns on invested capital and returns on equity (from Bloomberg, August 2008) for some leading solar PV companies.

Polysilicon/Wafer

Tokuyama: ROIC 23.18 percent, ROE 9.66 percent

Wacker: ROIC 50.46 percent, ROE 24.68 percent

MEMC: ROIC 55.87 percent, ROE 51.61 percent

Sumco: ROIC 50.42 percent, ROE 23.01 percent

Clearly, polysilicon/wafer companies on average have been generating excellent returns on capital as well as on equity. From a business standpoint, this arises from the shortfall of polysilicon capacity in recent years, compared to strong growth in demand from the solar PV sector. While these returns may revert back to normal levels once additional capacity comes online, this information does suggest to an investor that these companies are generating excellent returns in the near term.

Cells/Modules

First Solar: ROIC 19.73 percent, ROE 20.99 percent

Suntech Power: ROIC 18.14 percent, ROE 22.24 percent

Q-Cells AG: ROIC 16.49 percent, ROE 13.11 percent

JA Solar: ROIC 21.63 percent, ROE 18.41 percent

Cell and module companies have also been generating fairly good returns, though not at the level of polysilicon/wafer companies. On a business level, this is driven by robust demand due to solar-friendly policies in many countries, especially in Europe.

Vertically Integrated

Trina Solar: ROIC 15.67 percent, ROE 13.62 percent

SunPower: ROIC 12.10 percent, ROE 1.36 percent

Renewable Energy Corporation: ROIC 23.12 percent, ROE 11.91 percent

Solarworld AG: ROIC 24.12 percent, ROE 17.57 percent

Solar Capital Equipment Manufacturing

Amtech Systems: ROIC 8.87 percent, ROE 8.98 percent

Spire: Negative returns

Polysilicon Companies

In recent years several companies have announced plans to start or expand polysilicon manufacturing. This has been a response to the polysilicon supply crunch over the last few years. How does an investor identify quality polysilicon companies? Here are a few aspects that investors should explore as supplements to valuations for polysilicon companies.

Does the company have a track record in building and ramping capacity?

This is critical because polysilicon manufacturing is a highly intricate, complex process with numerous obstacles to be overcome. A company like Hemlock Semiconductor, which has a history of successful execution, is more likely to successfully execute planned capacity additions than a company that doesn't have similar experience. The company's attention to quality could also signal useful information to an investor. Again, to take the example of Hemlock, the company adheres to the ISO 9001:2000 revised standards. On the other hand, there are several companies that do not have an extensive track record in polysilicon manufacturing. Therefore, investors would be more cautious when evaluating their capacity expansion plans, and also in assessing how these plans affect valuation.

Polysilicon manufacturing is highly capital intensive, which means large additions of capacity require proportionately large amounts of capital. For example, Asia Silicon (Qinghai) Co Ltd. gives a few indicative numbers for the capital expenditures involved in setting up silicon capacity in different geographies (A New Silicon Supplier's Role in the Solar Electricity Value Chain," presentation by Asia Silicon, slide 9):

- Partial Siemens-based polysilicon plant with a U.S.-based gas partner: $220 million for a capacity of 1,550 metric tons per year
- New Siemens-based plant in the United States: $250 million for a capacity of 1,500 metric tons per year
- Upgraded metallurgical silicon-based plant in Europe: $450 million for a capacity of 5,000 metric tons per year
- Full Siemens-based polysilicon plant in China: $130 to $175 million for a capacity of 1,500 metric tons per year

An investor would therefore want to know whether a poly company has the requisite funds to actually build the kind of capacity it is planning. In this regard, prepayments from customers could help a company finance capacity additions. By knowing the amount of prepayments received in relation to the total amount of funding required, an investor could get a fair idea about the financial viability of a polysilicon project. For example, by September 2008, Hoku Scientific had ~$270 million in prepayments to fund its $390 million polysilicon plant.

INVESTMENT PROSPECTS
FOR SOLAR PV STOCKS

Of the many solar companies available to the public equity investor, we have selected some of the most prominent names and raise a few concerns specific to each one, as of 2008. The list does not imply that these companies are superior or inferior to other solar companies not listed. Moreover, as the solar markets evolve, the issues relevant to estimating each company's earnings drivers and cost of capital may vary.

MEMC

Given that poly supplies are likely to ease by the second half of 2009, MEMC's long-term picture is not clear. Indeed, prospects for MEMC would depend on whether poly off-take would remain strong enough to sustain the massive amounts of polysilicon capacity coming online in the next two to three years. The situation is further complicated by the emergence of upgraded metallurgical silicon as a potentially viable alternative to Siemens-based polysilicon.

LDK Solar

LDK Solar has announced plans for setting up polysilicon capacity on a large scale, with 16,000 MT planned by the end of 2009. This means LDK will eventually become a vertically integrated player with operations spanning silicon to wafering. The poly produced internally is meant to feed a massive capacity expansion plan, with 3.2 GW of wafers planned by the end of 2010. Over the long term, stock price performance would be contingent on the extent to which poly operations succeed and also on the company's ability to simultaneously ramp up wafering capacity on the large scale it envisions.

One important aspect of LDK Solar's strategy is articulated in its 20-F filing with the SEC for FY07. LDK has stated that its positioning as a pure wafer/poly manufacturer means it does not have any conflicts of interest with its customers who are cell/module manufacturers. According to the company, this positioning means it is easier to enter into long-term strategic relationships with its customers and also to obtain feedback. In a more general sense, we may note that a solar PV company's plans for vertical integration also need to take into account the possibility of any conflicts of interest with customers.

JA Solar

China-based JA Solar has one of the lowest nonsilicon costs in the solar PV industry. Indeed, JA Solar's processing costs are just 20 cents per watt. Its capital costs, at 25 to 30 cents per watt, are also low. With the shortage of poly supplies potentially easing by the second half of 2009, JA Solar could see the benefits of its low nonsilicon costs stand out even more, under a normalized silicon price scenario. To summarize, the investment thesis for JA Solar is based on a pure cell play with extremely low manufacturing and capital costs.

Solarfun Power

Solarfun Power has plans to ramp module capacity and also to integrate backward into silicon. As we explain later in the book, Solarfun's backward integration strategy is fraught with several risks that an investor might want to consider. Given these risks, positive developments on the poly front could potentially act as catalysts for the stock price.

Suntech Power

With demonstrated capability in ramping up capacity and an aggressive raw material procurement strategy, Suntech Power represents a quality company. Recent moves by Suntech to augment its silicon supplies point to a concerted effort to scale up operations. For example, the company recently signed an agreement to buy as much as 220 megawatts of silicon wafers from Wacker-Schott, a joint venture between Wacker and Schott Solar. It has also acquired stakes in poly manufacturers (such as Nitol).

Key aspects of Suntech Power's investment thesis include its economies of scale, which can give it a per-watt cost advantage

over competitors since fixed costs are spread over a larger megawatt capacity.

SunPower

SunPower's investment thesis is based on the company being the
manufacturer of the highest efficiency panels commercially available in the solar market. Its forward integration into the installations
business with its acquisition of PowerLight, is a play by SunPower
on the systems integration space as well. Further, the company also
seems poised to create a substantial presence in the financing of
large scale systems, as was seen in its recent agreement with PG&E
involving 250 MW of solar PV.

Large scale contracts such as the one with PG&E could act as
catalysts for the stock. Such announcements could also have broader
positive implications for the stock, in terms of validating emerging
business models (for example, PV manufacturers also financing the
system) in the eyes of investors.

Canadian Solar

Over the last few quarters, Canadian Solar has been increasing its
focus on upgraded metallurgical silicon, with production yields and
efficiencies using this route demonstrating an encouraging trend.
The fact that other major companies, such as Q-Cells, have decided
to adopt an upgraded metallurgical silicon strategy bolsters the
case for this technology. The investment thesis for Canadian Solar
would be based on upgraded metallurgical silicon taking off in line
with the company's expectations.

Evergreen Solar

Evergreen Solar's investment thesis is based on the fact that it has
the technology to produce solar cells with the least silicon consumption, at less than five grams per watt. As we explain in Chapter 6,
even under a normalized silicon price scenario, Evergreen Solar
would have a cost advantage of roughly 10 percent due to its
lower silicon consumption based on its proprietary string ribbon

technology, which will be a major advantage as the industry becomes increasingly commoditized.

Yingli Green Energy

Yingli Green is a vertically integrated player with a presence from silicon to module manufacturing. With low capital costs of ~$1/W (ingot to module manufacturing included), Yingli can ramp capacity in a cost effective manner. The investment thesis would therefore be related to the fact that Yingli Green Energy is a low cost, vertically integrated player in the solar PV industry.

Trina Solar

Trina Solar is also a vertically integrated company, with operations ranging from ingot/wafer to modules. With low capital costs of $1/W (covering all steps from making ingots to modules), Trina Solar also has the capability to ramp capacity at relatively low costs.

Energy Conversion Devices

Energy Conversion Devices occupies a unique position in the building integrated photovoltaics (BIPV) space, with its flexible amorphous-silicon-on-stainless-steel product. Any news on subsidies related to BIPV could act as a catalyst for its stock.

First Solar

First Solar appears poised to be the first commercial solar company to reach grid parity. With its CdTe technology and low cost manufacturing enabling 50 percent-plus gross margins and a clear road map to grid parity, First Solar could capitalize on the surging global demand for electricity once it becomes cost competitive with other sources of electricity. In the short term, contracts with large utilities could act as catalysts for the stock. In the long run, however, the company's reliance on scarce tellurium for its CdTe solar technology could cause growth to stall, costs to rise, and the company's stock price to fall.

CHAPTER 5

Volatility and Risk

Valuation tools such as those described in the last chapter help investors decide the price they're willing to pay for a particular solar stock. But once you've used valuation methods to set a target price for a stock, what tools can you use to assess where that stock's price will go? That crucial step is the subject of this chapter, which addresses the volatility and risk factors that may cause a stock's price to rise or fall.

Part of the assessment process depends on your investment philosophy. Warren Buffet and many other successful investors look at a company's long-term prospects and are willing to be patient and endure short-term losses in the belief that the stock will increase over time. Many other investors, however, seek to maximize returns by getting in and out of stocks quickly. There is certainly a great deal of profit to be made from following such a "fast money" trading philosophy in solar stocks, but investors must know how to recognize the key factors that indicate a particular solar stock is "in play." Long-term investors, on the other hand, need to know how to resist getting fooled and how to use knowledge of short-term dynamics to their advantage by adding or reducing their holdings during periods of excess volatility. Both groups can achieve their investment goals by understanding catalysts that have caused solar stock price movement in the past.

This chapter examines how these catalysts have correlated with the short-term performance of solar stock prices in the past, and how investors can gain an idea of what types of events could potentially move stocks in the future. We'll look at actual trades from the solar stock sector and analyze how they were driven by catalytic events.

CAPACITY EXPANSIONS AND PRODUCTION GROWTH

News of capacity expansion and production growth signifies a potential increase in revenue in the future, contingent on the company's ability to execute and obtain the necessary raw materials and equipment. The importance of this type of catalyst depends on the scale of the planned capacity expansion, the company's track record in execution, its perceived ability to source raw materials for the planned expansion, and the timeline for the new capacity to come online. The potential for increased revenue translating into higher cash flows and share valuation can cause a company's stock price to surge.

On the other hand, there may be times when a company's decision to *cancel* previously announced capacity additions acts as a positive catalyst. This might sound counterintuitive, but such an announcement may reduce volatility and therefore the required rate of return on the stock, sending valuations up.

A key question here is: When can a trader expect news of a rollback on capacity expansion or vertical integration plans to send a company's stock price up? In many cases, the answer is that if a company does not have a great track record and the market is skeptical about its capacity expansion plans, rolling back the decision can reduce stock volatility and indeed push the stock price up. This principle can be used as a basis for short-term trades.

RAW MATERIAL SUPPLY CONTRACTS

For crystalline silicon-based PV, the short supply of polysilicon in recent years has meant that securing supplies was considered crucial for ensuring revenue growth. Indeed, capacity and production growth require not only the setting up of plants and machinery, but also ensuring that there is adequate feedstock to support this growth. Despite the shortsighted nature of many take-or-pay contracts and the possibility that these contracts were being struck at prices too high to grow profitably in the future, during much of 2006, 2007, and 2008 a polysilicon supply contract often moved a stock up considerably, demonstrating that the short-term and long-term often contradict.

Traders should, therefore, be on the lookout for any news of favorable supply contracts for polysilicon or silicon wafers because it immediately increases the feasibility of any capacity expansion plans. However, as the downturn currently underway in solar persists, existing contracts that are uneconomical today may be renegotiated or broken which very well may be the positive stock catalyst going forward for a time. In a sense, the trade in this case is based on greater certainty in revenues, cash flows, and profitability.

On the other hand, for many companies based on thin film PV, raw material supply may be plentiful; therefore, any news on raw material agreements may turn out to be a nonevent as far as the markets are concerned. However, a longer-term risk for the thin film companies may be based on the supply of a key raw material (such as the active ingredient in cadmium telluride or CIGS/CIS) becoming constrained. A situation may arise where news of raw material supply contracts for thin film players may be greeted with extreme prejudice by investors while simultaneously creating enthusiasm for crystalline-silicon-based companies.

In general, the importance of this catalyst depends on how the supply of the relevant raw material is constrained (and therefore the extent to which uncertainty is mitigated by the announcement of the contract). In recent years this factor has proved a major catalyst for crystalline-silicon-based cell and module companies and not so much for thin-film-based players.

REVENUE VISIBILITY

Announcements of sales contracts won by a company can often act as a powerful short-term positive catalyst on its stock price. In fact, as we shall see in one example, even if a company's fundamentals are not very strong (for example, even if a company does not have an adequate track record in execution), news on revenue visibility can send its stock price sharply upward. This catalyst drives trades on the basis of increased revenues and increased cash flows. Also, if a company signs a prepayment supply contract (which has been quite common for polysilicon manufacturers) and the short interest is through the roof, traders can expect a short squeeze. However, given the oversupply in polysilicon and throughout the supply chain, renegotiated and broken supply contracts will have different influences on the stocks depending on whether the changes are perceived as positive or negative for sales and earnings.

Revenue visibility is a very important catalyst in the solar PV industry, regardless of whether a company is based on crystalline silicon or thin film, or even where a company is in the supply chain.

NEWS ON GOVERNMENT POLICIES

As the solar industry is to a large extent currently dependent on government policies and subsidies, any news in this regard could have a potential effect on solar stock prices.

As we described in Chapter 2, the global nature of the solar industry means that news on solar-related government policy in Spain could have an impact on companies based in China or the United States. The details of whether and to what extent news in a country could affect a company's stock price depend on various factors, such as these:

- If a country decides to reduce solar subsidies, it would have a strongly detrimental effect on companies whose revenues are heavily exposed to that country and a milder effect on companies with little or no exposure to that country. Although the global nature of the solar PV industry means that affected companies can possibly shift sales to other countries, establishing new distribution channels and customer relationships to compensate for lost sales, this occurs only with a time lag, if at all. This could push revenues and cash flows further out into the future, bringing down the valuation.
- If a country announces a major push into solar PV, it could have a positive impact on companies because it presents an expanded market where any company can potentially

come in and achieve increased sales. However, stock prices of companies that already have operations in that country could stand to gain much more in the near term as the markets factor in their greater ability to gain a foothold on the newly expanded market.

- The importance of news on government policy also depends on the position of the company in the supply chain. The operations of system integrators/installers are much more localized than those of players further upstream. Therefore, if an installer with operations only in Germany suddenly finds that the subsidy picture has become unfavorable, its stock price could be heavily affected because it would be extremely difficult and time-consuming to make up for lost revenues.

NEWS ON ETHICS/CORPORATE GOVERNANCE ISSUES

News on corporate governance or ethical issues can be a catalyst for a trader to sell or short the stock. In the solar PV industry, given the limited operating history of many companies, news (or even rumors) of corporate-governance-related issues can have a particularly negative impact. This catalyst is independent of whether the company is based on crystalline silicon or thin film, or its position in the supply chain. Rather, its importance depends on the significance of the reported ethics/corporate governance issue.

A REPUTABLE INVESTOR INCREASING STAKE/INSIDER OWNERSHIP

If a reputable investor buys a significant number of shares in a solar PV company, it could send positive signals to the market, resulting in an increase in stock price. If a company has only recently started its operations, the market may not feel completely confident about its prospects. However, if a reputable investor picks up a large stake in the company, the market could interpret it as a positive signal under the belief that the reputable investor must have bought the stock only after extensive due diligence. The importance of this catalyst usually increases in the case of a company that has been listed only recently and whose operations remain particularly unclear to the market.

Similarly, if an insider increases his or her stake in a company, it could act as a positive catalyst, especially for companies that have only recently started operations.

APPROACHING PREPAYMENT DEADLINES

Sometimes companies may have supply agreements that stipulate that customers make prepayments upon reaching certain milestones. Indeed, this kind of agreement is very common with polysilicon companies, whose plants may be funded significantly through such agreements.

Traders obtain valuable information on a company's execution based on whether these prepayments occur or not. These prepayment deadlines are therefore a catalyst for stock prices.

The importance of this catalyst depends on whether such prepayment agreements are prevalent, given the company's position in the supply chain, the amount of prepayment involved, and the significance of the milestone.

VOLATILITY ANALYSIS AND CLASSIFICATION OF CATALYSTS: EXAMPLES

Having highlighted the principles and logic behind various catalysts, we will now consider some actual examples of such catalysts as they apply to various solar companies. We have organized these examples by their primary business (PV companies, solar equipment manufacturers, etc.) and the positions of various companies within the supply chain.

Polysilicon/Ingot Companies

Approaching Prepayment Deadlines
Deadlines related to prepayment contracts with customers can act as a catalyst for stock prices, as illustrated by the example of Hoku Scientific.

On September 4, 2008, Hoku Scientific announced that it had terminated its poly supply agreements with GEWD (a 100 percent subsidiary of Solar Fabrik AG) and Sanyo due to oversubscription at Hoku's planned 3,500 MT plant in Pocatello, Idaho. These terminations allowed Hoku to reallocate part of the expected production from the Pocatello plant to Kinko Energy, Tianwei New Energy, and Solargiga, and also possibly to achieve more favorable sales agreements using its current expected capacity.

As a result, Hoku must return prepayments of $2 million each to GEWD and Sanyo. However, with its other sales contracts, Hoku had $270 million in prepayments from customers to finance the $390 million Pocatello plant (for the remaining $120 million required in financing, Hoku is contributing $47 million, and potential new customers are contributing $73 million). Of that $270 million in prepayments, not all have been received.

This suggests potential catalysts in the form of approaching prepayment deadlines. In particular, Hoku Scientific has the following cash infusions coming up in the near future in the form of prepayments from its various sales contracts, subject to the company reaching certain milestones:

- $15 million each in Q1 CY09, Q2 CY09, and Q3 CY09 from Suntech Power
- $20 million in Q1 CY09 and $5 million in Q1 CY10 from Solarfun
- $25 million in Q1 CY09 from Kinko
- $10 million in Q1 CY09 and $5 million in Q1 CY10 from Tianwei
- $20 million in Q1 CY09 and $5 million in Q1 CY10 from Solargiga

In all, Hoku Scientific will be receiving $90 million in Q1 CY09, $15 million in Q2 CY09, $15 million in Q3 CY09, and $15 million in Q1 CY10 if it reaches certain construction and shipping milestones as planned. Announcements of prepayments occurring as scheduled could be a positive catalyst since they indicate that the company's execution is on track. However, if the company misses any targets and prepayments do not come in as a result, the stock price could be driven down in the market, at least in the near term.

In general, the tightness in polysilicon supplies in recent years has meant that PV companies have been willing to fund the expansions of poly companies in the form of prepayments, often with milestone clauses attached, as in the case of Hoku Scientific. These can act as positive or negative catalysts, depending on whether the prepayments come in as scheduled.

Revenue Visibility

News on revenue visibility can boost short-term stock prices, sometimes regardless of the underlying fundamentals. This can be clearly seen in the case of a news event that sent Hoku Scientific's stock price soaring.

On July 30, 2008, Hoku Scientific announced that it had entered into a contract with Jiangxi Kinko Energy Company, Ltd., to supply polysilicon worth $298 million over a 10-year period. To put this in context, Hoku is a troubled company with a history of failures. Indeed, the company initially started out in the fuel cell business, and when it failed at that business, decided to go into the polysilicon manufacturing business. The prospects of Hoku's success in the polysilicon business are by no means certain. Despite this, its stock price shot up ~17 percent upon the announcement. Some of this movement can be attributed to the short interest, but clearly the market values long-term sales contracts because of their positive effects on revenue visibility.

Wafer Companies

News on Ethics/Corporate Governance Issues

As one would expect, any significant ethical or legal issues coming to light are likely to have an adverse impact on stock prices.

On October 3, 2007, LDK Solar fell by more than 24 percent on news that its financial controller had left the company, alleging poor controls and discrepancies in the accounting of polysilicon inventory. While such events cannot be completely predicted, it may at times be possible to obtain a broad indication of corporate governance issues in a company. Again, to take the example of LDK Solar, its public accounting firm had raised red flags about the financial controls while auditing the FY06 results. Such indications could be key decision factors.

On December 17 2007, LDK Solar's American Depositary Receipts (ADR) rose nearly 20 percent on news that the audit committee (which included independent experts) had found no "material errors" in the accounting of polysilicon inventory.

Revenue Visibility
On December 10, 2007, LDK Solar announced that it had entered into a "take or pay" contract with German solar cell producer Q-Cells AG to supply more than six GW of wafers over a 10-year period starting in 2009. The stock price went up 28.6 percent in response to news of this significant development. Similarly, on March 13, 2008, when LDK announced that it had sold out 2008 output and ~90 percent of 2009 output, the stock price responded with a 23.7 percent rally in a single day.

Cell and Module Companies

Capacity Expansion and Production Growth
On April 27, 2007, XsunX announced that it was planning to build a production facility for building-integrated photovoltaics in the United States. The stock price went up by over 23 percent that day.

Revenue Visibility
On October 16, 2006, Evergreen Solar announced that it had reached an agreement with a company called Mainstream Energy to supply solar modules worth $100 million over a four-year period. This news was followed by a stock price appreciation of over 15 percent that day.

On June 19, 2008, Evergreen Solar's stock price went up by ~20 percent on news that it had secured a contract to sell $600 million of its solar panels to groSolar and Wagner & Co Solartechnik GmbH over a four-year period, bringing the total order backlog to $1.7 billion.

Raw Material Supply Contracts

This section is based on a research note published by Alternative Energy Investing (AEI).

On Friday, March 17, 2006, Evergreen Solar's stock price declined by nearly 11 percent. Just a day earlier, 10 minutes before the close of trading on Thursday afternoon, Evergreen Solar had filed its Form 10-K annual report. The filing revealed that the company's sole source supplier, MEMC (the only maker of high-purity granular solar polysilicon) had exercised an option to cancel its supply agreement and pay an excess fee in lieu of shipping 53 MT remaining on a 100 MT polysilicon supply agreement.

This meant that Evergreen Solar now had to seek help from Renewable Energy Corp. in the form of accelerating a recently negotiated long-term contract for granular polysilicon that Evergreen had not anticipated needing for some time. At the time, Renewable Energy had not yet started commercial production of granular polysilicon, so supplies in the near term were far from assured.

The Evergreen Solar string ribbon process requires granular polysilicon. Ultrahigh purity granular polysilicon, of the specification made by MEMC, is desirable for this process since it can be easily melted to top off the crystal growing crucible, allowing a longer silicon ingot crystal without the need to shut down the furnace. However, knowing that it would no longer have access to granular polysilicon from MEMC, Evergreen Solar decided to source polysilicon in chunk form until the granular form became available. This would have to be crushed into tiny granules in order to make it usable.

This presented several complications. First, Evergreen Solar had never before used crushed chunk polysilicon for production on a commercial scale, so investors could not feel assured that this would be successful. Further, the additional impurities existing in chunk alone could cause a material decline in yields. Moreover, if the crushing process contributed to an increase in impurity levels, yields could theoretically collapse. Also, the news meant a potential delay in transitioning to the "thin ribbon" technology, which added more risk to consensus numbers.

The sharp fall in Evergreen Solar's stock price was a response to these significant risks faced by the company at the time.

On October 9 2007, China Sunergy reached a supply agreement with Luoyang Zhonggui High-tech, a Chinese polysilicon-manufacturing company, for the supply of an aggregate of 106 metric tons of virgin polysilicon for 2007 and 2008. Following this, the Nasdaq-listed American Depositary Shares (ADS) rallied 20.7 percent, a reflection of the importance placed by the market on assured poly supplies.

On October 25, 2007, Suntech Power rose by over 16 percent on news that it had signed an agreement with Asia Silicon for the procurement of $1.5 billion of polysilicon from its China-based plant, over a period of seven years.

News of a Reputable Investor Increasing Stake/Insider Trading

The purchase or sale of shares by insiders can have an influence on stock prices. For example, on October 18, 2007, Ascent Solar was up by over 16 percent. This was preceded by news of the purchase of ~$906,000 worth of shares by a director in the company. To put this into context, the average purchase price for this was $15.64 and the market price at close on October 17 was $17.22.

This sent positive signals to the market about the company's prospects and contributed to the sharp increase in the stock price to $20 that day.

In the short term, actions by reputable investors could be an important driver of prices. Similarly, news of a respected player in the industry picking up a strategic stake in a company could boost prices, at least temporarily.

For example, on December 3, 2007, Good Energies Investments BV, a reputed Swiss investor focused on the renewable energy sector, agreed to increase its stake in Solarfun Power from 6.3 to 34.7 percent. Following this news, the stock price rose sharply, by ~26 percent on that day and ~27 percent the next day. This suggests that a significant investment by a well-respected firm could be a catalyst for short-term price movements.

News of Governmental Programs/ Subsidies/Incentives

Given that the solar industry is dependent on government policies and subsidies, any news in this regard could have an effect on solar stock prices.

For instance, on December 26, 2007, prices of most solar stocks rose on reports that the Chinese government was planning a major push into renewables. China Sunergy in particular witnessed a strong 39 percent rally on that day (followed by another 16-plus percent gain the next day), with other stocks such as Solar Enertech (+34.8 percent), Solarfun Power ADR (+8.5 percent), and Canadian Solar (+7.2 percent) posting gains as well. This underscores the significant impact that news of government policy changes could have on solar stocks.

Vertically Integrated

Capacity Expansion and Production Growth

On April 14, 2008, Trina Solar announced that it was canceling plans to develop a $1 billion polysilicon plant with a capacity of 10,000 MT/yr, because supply conditions had made it more favorable to source the raw material externally. This meant Trina Solar would no longer be pursuing an expensive project with complex manufacturing requirements in an area where it lacked expertise. The market rewarded this strategic move immediately, sending the ADR price up 15.7 percent that day.

Given the relatively complex nature of setting up and running manufacturing operations in the solar industry, any news related to this area is closely tracked and acted upon. Conversely, news of problems related to production ramp-up or any disruptions in routine operations can drive down stock prices in the near term.

Solar Equipment

Revenue Visibility

On May 15, 2007, XsunX signed an agreement with Lambda Energia to supply 25 MW of production equipment worth more than $41 million. The stock price shot up by nearly 23 percent that day in response to this news.

On July 11, 2007, Amtech Systems announced that it had secured $4.4 million in new orders for its diffusion processing systems from solar cell manufacturing customers. The news sent the company's stock soaring 20.7 percent.

On the other hand, cancellations or delays in shipments could have an adverse impact on the stock price. For example, on December 13, 2007, Amtech Systems announced that some of its shipments (value of delayed shipments: $2 million; Amtech's net revenues for 2Q07: $10.5 million) would be delayed because a customer had asked it to shift the delivery from Europe to Asia. The single-day hit on the stock price amounted to more than 25 percent.

News of a Reputable Investor Increasing Stake/Insider Trading

On March 27, 2008, Ascent Solar's stock surged by over 17 percent on news that Norsk Hydro had exercised an option to purchase 2.3 million additional shares of Ascent, increasing its total stake to 35 percent.

Other Companies Reporting Results

News on other companies could provide valuable insights into how the solar industry is poised, and could drive price changes in a wide range of stocks.

On November 14, 2007, several U.S.-traded solar stocks surged as a result of upbeat results and forecasts at three German companies, namely, Q-Cells, Phoenix Solar, and Aleo Solar. The upward direction cut across many parts of the value chain. For example, on that day, SunPower, which produces and installs cells and modules, was up 6.9 percent, and Akeena Solar, which is a pure-play installer, was up 9.1 percent. The general idea here is that there might be instances where certain companies may report results that prompt investors to revise their outlook on the broader industry, and this could impact prices.

Earnings Reports and Analyst Actions

There are numerous instances of earnings reports and analyst actions having a major short-term impact on stock prices.

For example, on November 21, 2007, Trina Solar saw a
~24 percent drop in its ADR price on an earnings miss: EPS came
in at 29 cents/ADR against consensus estimates of 48 cents/ADR.
As another example, First Solar is a stock that has seen its price
boosted by earnings surprises on several occasions. For instance, on
February 14, 2007, November 8, 2007, and February 13, 2008, First
Solar saw sharp increases (27.5, 34.3, and 30.1 percent, respectively)
in its stock price after earnings releases. Similarly, analyst upgrades
can also have a significant impact on short-term prices.

However, while traders with a short-term agenda can certainly
anticipate volatility in the light of such events, it is difficult to use
these as reliable, predictive catalysts for trades.

Issuance of Securities

On June 26, 2008, China Sunergy announced that it had completed
the pricing of $50 million of convertible senior notes due 2013. These
notes, which would pay interest at 4.75 percent per annum, would
be convertible to American depository shares at an initial price of
$12.30 per ADS. This news was viewed unfavorably by the market,
sending the ADS price down by over 16 percent that day.

MEASURING VOLATILITY

The average absolute percentage changes in stock prices for U.S.-
traded solar companies give an indication of the volatility of different solar stocks. Specifically, these numbers represent the average daily percentage *changes* in stock price, without distinguishing
between positive and negative changes. The numbers in Table 5.1
cover data from January 2006 (or the date of listing for companies
which became listed after January 2006) through July 2008.

The stocks at the beginning of the list are prone to relatively
large daily fluctuations, and as we go down this list, we find stocks
that are relatively stable in terms of daily fluctuations.

PUT-TO-CALL OPEN INTEREST RATIO

The ratio of open interest in put options to call options in a particular stock can be a useful indicator of the sentiment in the market for

T A B L E   5.1

Average of Absolute Daily Percentage Changes for U.S.-Traded Solar Stocks

Company Name	Average Absolute % Changes	Company Name	Average Absolute % Changes
Renesola Ltd-ADR	5.77	Solar Enertech Corp	3.91
China Sunergy Co Ltd-ADR	4.70	First Solar Inc	3.24
LDK Solar Co Ltd-ADR	4.70	Emcore Corp	3.17
Open Energy Corp	4.56	Suntech Power	2.94
Yingli Green Energy	4.48	Evergreen Solar Inc	2.90
XsunX Inc	4.46	SunPower Corp	2.79
Solarfun Power	4.45	BTU International Inc	2.70
Akeena Solar Inc	4.38	Energy Conversion Devices	2.68
Ascent Solar Technologies	4.29	Spire Corp	2.55
Trina Solar	4.25	MEMC	2.47
Worldwater & Solar Tech	4.18	Amtech Systems Inc	2.46
Daystar Technologies Inc	4.02	Aixtron AG-ADR	2.37
JA Solar	3.98	Applied Materials Inc	1.41
Canadian Solar Inc	3.96		

a stock. If the put-to-call open interest ratio is more than one, it indicates that general sentiment in the market is negative. In general, the higher this ratio is, the more negative the market's view of the stock. An investor can use this as an input when deciding whether and when to invest in a solar stock.

THE ECONOMICS OF FABRICATION EXPANSION

When a solar company announces that it plans to expand capacity, investors naturally want to know how this will impact the company's stock price. Understanding the specific economics of capacity expansion is the key to knowing how much an announced plan is going to cost and how it will affect the company's bottom line.

In the terminology of the industry, a factory that manufactures semiconductor devices is known as a fabrication plant, or "fab" for short. In general, increased line capacity at fabs results in better economies of scale on a per-module basis. Simply put, the more modules a fab line can produce, the better the margins. According to Jeannine Sargent, CEO of Oerlikon Solar, 20 to 40 MW module capacity is relatively small, 60 to 80 MW is the current scale ordered, and 120 MW lines are now being installed. Jeannine is of the view that even greater capacity would come from multiple 120 MW lines that could be distributed around the world, with each producing ~4100 modules/day at 1.3 m^2.

The Need for PV Capacity Expansions and Scale

The demand for solar PV has pushed many leading players in PV manufacturing, as well as many new entrants, to ramp up production aggressively. Continued robust demand is leading to better economies of scale in production, which in combination with a steady improvement in conversion efficiencies, is helping the solar PV industry to attain a cost-per-watt level that is closer to the ultimate goal of reaching grid parity at peak usage levels.

The solar PV industry is steadily shifting toward large scale manufacturing facilities in order to capitalize on the economic scaling. As mentioned in the previous section, the vast majority of fabrication facilities in the solar PV industry are small and below 25 MW of annual production levels. Just increasing the number of such conventional units in a bid to reach the grid parity might prove to be insufficient to achieve the required cost-per-watt reductions, hence the need for huge capacity expansions. This is the reason many conventional crystalline cell and the emerging thin film producers are expanding their manufacturing units.

MEMC is adding to its polysilicon capacity at its Texas facility. Renewable Energy Corp. has also announced its plans to set up a manufacturing unit in Canada, where it will invest $1.2 billion. The construction is expected to begin in 2010, and the phase one production capacity will be available from 2012 on.

The thin film manufacturing area is also gaining considerable interest from new entrants. Some major equipment suppliers are

Applied Materials, Oerlikon Solar, and Ulvac. These companies will be able to overcome the polysilicon supply shortages and high raw material costs associated with supply constraints. Thin film has provided these companies the ability (depending on the capital they have) to ramp the production to larger scales without the silicon shortage pulling them back.

One example is First Solar, which expects to reach about one GW scale by the end of 2009 as it continues the ramp of all four of its plants in Malaysia at full swing. The capital spending in 2008 to achieve this capacity expansion goal is approximately $500 million. First Solar will also double the size of its production site in Ohio by 500,000 square feet. The company aims to lower the manufacturing cost per watt at a rate comparable to its lowest cost facility in Malaysia through economies of scale.

Another example from the thin film PV domain is AVA Solar. The company is setting up a production facility in Colorado that is expected to produce 200 MW of PV modules annually. Showa Shell Solar has plans to be the first company to have a one GW fab based on the CIS technology by the year 2011. The company has announced a 7 billion yen investment in order to increase its production capacity to 1,000 MW/1 GW by 2011 at its Miyazaki plant.

Case Analysis: Case 1

A close analysis of the economics of an individual fab operated by a typical solar company will help investors understand just how critical capacity expansion is.

Assume that a 300 MW fab has a life of 15 years. From the month the company announces either a new fab or expansion at an existing fab, it takes about eight quarters for the new or expanded fab to achieve full capacity. That timeline assumes that the construction starts in the quarter the announcement is made and takes about four quarters for completion. The estimated capital expenditure (capex) cost for construction is $300 mm, based on a capex per watt of ~$1.00. Also, from the moment when construction is completed, we assume that it will take three quarters for capacity to get fully ramped up, during which the company will expend an additional ~$90 mm as ramp cost.

Now, we can estimate the NPV (net present value) of the new 300 MW fab, based on the following assumptions:

- Capex/watt: $1/watt
- Capacity utilization: 80 percent (0.8)
- Capex cost: capex/watt * MW* 0.8
- Revenue: ASP * MW
- ASP/watt: $4/watt (ASP is the average selling price.)
- Cost/watt: $3.2/watt

Using the data in Table 5.2 and calculating the same for 15 years, which is the life of the fab in this case, we can calculate the cash flows for all 15 years, and hence the NPV and the IRR.

In Table 5.3 we convert Table 5.2 into a yearly format where for the first year the fab life cycle is construction and for the second year it is ramp and production. Going forward from the third year to the fifteenth year it remains in production.

During construction the capex cost incurred is $300 mm, and it shows as a negative value/cash outflow in the cash flows. In the second year the cash flows are still negative but the value increases and the gross margin is 20 percent as the company will start production in quarter four of the second year. Third year onward till the fifteenth year, the company has positive cash flows and the gross margins remain in the range of 15 to 20 percent.

The values we obtain for NPV (net present value) and IRR are $128 mm and 20 percent, respectively. Hence, it is economical to expand manufacturing, as in the fifteenth year of production the total revenue is $174.81 mm, the gross profit is $32.2 mm, the gross profit margin is 18.4 percent, and there is positive cash flow of $64.4 mm.

Case Analysis: Case 2

Let us consider another case where there is a delay in the ramp.

Again, assume that a 300 MW new or expanded fab has a life of 15 years and takes about eight quarters from the month of announcement to achieve full capacity. The estimated capex cost for this case is $300 mm based on a capex per watt of ~$1.00. Also,

TABLE 5.2

Fab Economics: Case 1

Fab Economics

$ mm	Year 1				Year 2				Year 3			
	q1	q2	q3	q4	q1	q2	q3	q4	q1	q2	q3	q4
Fab life cycle	Con.	Con.	Con.	Con.	Ramp	Ramp	Ramp	Prod.	Prod.	Prod.	Prod.	Prod.
Capex cost	$30.00	$90.00	$90.00	$90.00								
Start-up cost					$30.00	$30.00	$30.00					
ASP/watt								$4.00	$3.68	$3.68	$3.68	$3.68
Total revenue								$80.00	$73.50	$73.50	$73.50	$73.50
Cost/watt								$3.20	$3.17	$3.17	$3.17	$3.17
Total cost								$64.00	$63.36	$63.36	$63.36	$63.36
Gross margin								$16.00	$10.14	$10.14	$10.14	$10.14
Cash flows	($30.00)	($90.00)	($90.00)	($90.00)	($30.00)	($30.00)	($30.00)	$32.00	$20.28	$20.28	$20.28	$20.28

Con. = Construction. Prod. = Production.

TABLE 5.3

Fab Economics: NPV and IRR Calculation

All Values in $ mm	Year 1	Year 2	Year 3	Years 4–12	Year 13	Year 14	Year 15
Fab life cycle	Construction	Ramp and prod.	Production	…	Production	Production	Production
Capex cost	$300.00						
Start-up cost		$90.00					
ASP/watt		$4.00	$3.92	…	$2.42	$2.30	$2.19
Total revenue		$80.00	$313.60	…	$193.69	$184.01	$174.81
Cost/watt		$3.20	$3.14	…	$2.02	$1.90	$1.78
Total cost		$64.00	$250.88	…	$161.39	$151.71	$142.60
Gross profit		$16.00	$62.72	…	$32.30	$32.30	$32.20
Gross margin		20.0%	20.0%	…	16.7%	17.6%	18.4%
Cash flows	($300.00)	($58.00)	$125.44	…	$64.61	$64.61	$64.41
NPV	$128.08						
IRR	20%						

we assume the ramp time to be three quarters, during which the company will expend ~$90 mm as ramp cost.

Now, we can estimate the NPV of the new 300 MW fab, based on the following assumptions:

- Capex/watt: $1/watt
- Capacity utilization: 80 percent
- Capex cost: capex/watt * MW* 0.8
- Revenue: ASP * MW
- ASP/watt: $4/watt (ASP = average selling price.)
- Cost/watt: $3.2/watt

But in this case, as depicted in Table 5.4, there is an extended ramp-up, and the fab takes more time than expected to become fully operational. The construction is completed in four quarters (or one year), and the ramp-up goes on for five quarters. The fab becomes fully operational in the second quarter of the third year.

Using the above data and calculating the same for 15 years, which is the life of the fab in this case, we can calculate the cash flows for all 15 years, and hence the NPV and the IRR.

For Table 5.5, we convert Table 5.4 into yearly format where for the first year the fab life cycle is construction, for the second year it is ramp, and for the third year it is ramp and production. Going forward from the fourth year to the fifteenth year, it remains in production.

During construction the capex cost incurred is $300 million, and it shows as a negative value/cash outflow in the cash flows. In the second year the cash flows are still negative, as the fab has the start-up cost of $120 million. From the third year on until the fifteenth year, the company has positive cash flows, and the gross margins remain in the range of 13 to 20 percent.

The values we obtain for NPV and IRR are about $44 million and 14 percent, respectively. Hence, we can easily conclude that even in this case it is economical to expand manufacturing, as in the fifteenth year of production the total revenue is about $178 million, the gross profit about $33 million, the gross profit margin more than 18 percent, and there is positive cash flow of about $66 million.

T A B L E 5.4

Fab Economics: Case 2

Fab Economics

$ mm	Year 1				Year 2				Year 3			
	q1	q2	q3	q4	q1	q2	q3	q4	q1	q2	q3	q4
Fab life cycle	Con.	Con.	Con.	Con.	Ramp	Ramp	Ramp	Ramp	Ramp	Prod.	Prod.	Prod.
Capex cost	$30.00	$90.00	$90.00	$90.00								
Start-up cost					$30.00	$30.00	$30.00	$30.00	$30.00			
ASP/watt										$4.00	$4.00	$4.00
Total revenue										$80.00	$80.00	$80.00
Cost/watt										$3.20	$3.20	$3.20
Total cost										$64.00	$64.00	$64.00
Gross margin										$16.00	$16.00	$16.00
Cash flows	($30.00)	($90.00)	($90.00)	($90.00)	($30.00)	($30.00)	($30.00)	($30.00)	($30.00)	$32.00	$32.00	$32.00

Con = Construction. Prod. = Production.

TABLE 5.5

Fab Economics (Case 2): NPV and IRR Calculation

All Values in $ mm	Year 1	Year 2	Year 3	Years 4–12	Year 13	Year 14	Year 15
Fab life cycle	Con.	Ramp	Ramp and prod.	…	Prod.	Prod.	Prod.
Capex cost	$300.00						
Start-up cost		$120.00	$30.00				
ASP/watt			$4.00	…	$2.47	$2.35	$2.23
Total revenue			$240.00	…	$197.65	$187.76	$178.38
Cost/watt			$3.20	…	$2.06	$1.94	$1.82
Total cost			$192.00	…	$164.68	$154.80	$145.51
Gross profit			$48.00	…	$32.96	$32.96	$32.86
Gross margin			20.0%	…	16.7%	17.6%	18.4%
Cash flows	($300.00)	($120.00)	$66.00	…	$65.93	$65.92	$65.72
NPV	$43.55						
IRR	14%						

Con. = Construction. Prod. = Production.

Case Analysis: Case 3

Let us consider another case where ramp is achieved ahead of time.

Again, assume that a 300 MW new or expanded fab has a life of 15 years and takes about eight quarters from the month of announcement to achieve full capacity. The estimated capex cost for this case is $300 mm, based on a capex per watt of ~$1.00. Also, we assume the ramp time to be three quarters, during which the company will expand ~$90 mm as ramp cost. Now, we can estimate the NPV of the new 300 MW fab based on the following assumptions:

- Capex/watt: $1/watt
- Capacity utilization: 80 percent
- Capex cost: capex/watt * MW* 0.8
- Revenue: ASP * MW
- ASP/watt: $4/watt (ASP is the average selling price.)
- Cost/watt: $3.2/watt

But in this case the ramp is achieved ahead of time, and the fab takes less time than expected to become fully operational. The construction is completed in three quarters, and the ramp-up is completed in another two months. The fab becomes fully operational in the second quarter of the second year itself, as can be seen in Table 5.6.

Using the above data and calculating the same for 15 years, which is the life of the plant in this case, we can calculate the cash flows for all 15 years, and hence the NPV and the IRR. In Table 5.7 we convert Table 5.6 into a yearly format where for the first year the plant life cycle is construction, for the second year it is ramp, and for the third year it is ramp and production. Going forward from the fourth year to the fifteenth year, it remains in production.

During construction the capex cost incurred is $240 million and it shows as a negative value/cash outflow in the cash flows. From the second year on till the fifteenth year the company has positive cash flows, and the gross margins remain in the range of 13 to 20 percent.

T A B L E 5.6

Fab Economics: Case 3

Fab Economics

$ mm	Year 1				Year 2				Year 3			
	q1	q2	q3	q4	q1	q2	q3	q4	q1	q2	q3	q4
Fab life cycle	Con.	Con.	Con.	Ramp	Ramp	Prod.	Prod.	Prod.	Prod.	Prod.	Prod.	Prod.
Capex cost	$30.00	$90.00	$90.00									
Start-up cost				$30.00	$30.00							
ASP/watt						$4.00	$4.00	$4.00	$3.92	$3.92	$3.92	$3.92
Total revenue						$80.00	$80.00	$80.00	$78.40	$78.40	$78.40	$78.40
Cost/watt						$3.20	$3.20	$3.20	$3.17	$3.17	$3.17	$3.17
Total cost						$64.00	$64.00	$64.00	$63.36	$63.36	$63.36	$63.36
Gross margin						$16.00	$16.00	$16.00	$15.04	$15.04	$15.04	$15.04
Cash flows	($30.00)	($90.00)	($90.00)	($30.00)	($30.00)	$32.00	$32.00	$32.00	$30.08	$30.08	$30.08	$30.08

Con. = Construction. Prod. = Production.

159

TABLE 5.7

Fab Economics (Case 3): NPV and IRR Calculation

All Values in $ mm	Year 1	Year 2	Year 3	Years 4–12	Year 13	Year 14	Year 15
Fab life cycle	Con.	Ramp	Ramp and prod.	...	Prod.	Prod.	Prod.
Capex cost	$210.00						
Start-up cost	$30.00	$30.00					
ASP/watt		$4.00	$3.92	...	$2.42	$2.30	$2.19
Total revenue		$240.00	$313.60	...	$193.69	$184.01	$174.81
Cost/watt		$3.20	$3.14	...	$2.02	$1.90	$1.78
Total cost		$192.00	$250.88	...	$161.39	$151.71	$142.60
Gross profit		$48.00	$62.72	...	$32.30	$32.30	$32.20
Gross margin		20.0%	20.0%	...	16.7%	17.6%	18.4%
Cash flows	($300.00)	$66.00	$125.44	...	$64.61	$64.61	$64.41
NPV	$43.55						
IRR	14%						

Con. = Construction. Prod. = Production.

The values we obtain for NPV and IRR are more than $280 million and 38 percent, respectively. Hence, even in this case it is economical to expand manufacturing, as in the fifteenth year of production the total revenue is about $175 million, the gross profit about $32 million, the gross profit margin more than 18 percent, and there is positive cash flow of about $64 million.

Conclusion for the Three Cases

After analyzing the three cases above, we can conclude that capacity expansions and scaling are profitable for solar PV companies in most scenarios.

THE FUTURE OF FAB EQUIPMENT SPENDING

Alternative Energy Investing estimates the production capacity for wafer-based solar cells was expected to increase by more than 60 percent, to greater than 12 GW in 2008. Table 5.8 shows the production capacity for cell (module) companies.

T A B L E 5.8

Global Production Capacity of Silicon Crystalline Cells

Cell (Module) Company	2005	2006	2007	2008
Advent Solar		10 MW	25 MW	25 MW
Alpex Exports				15 MW
Arise Technologies Deutschland GmbH			30 MW	115 MW
AZUR Space Solar			15 MW	15 MW
Bharat Heavy Electricals Limited (BHEL)	Nil.	5 MW	20 MW	50 MW
BP Solar	147 MW	295 MW	335 MW	335 MW
Canadian Solar Inc		25 MW	150 MW	250 MW
Changzhou Eging			250 MW	250 MW
China Sunenergy	100 MW	190 MW	300 MW	350 MW
Chint Solar		25 MW	100 MW	100 MW
Conergy AG		50 MW	275 MW	275 MW

TABLE 5.8

(Continued)

Cell (Module) Company	2005	2006	2007	2008
DelSolar (Delta Electronics)	50 MW	50 MW	100 MW	100 MW
E-Ton Solar	35 MW	150 MW	200 MW	320 MW
Eoplly		20 MW	40 MW	60 MW
ErSol	60 MW	210 MW	300 MW	390 MW
E-Ton Solar	38 MW	38 MW	38 MW	38 MW
Evergreen Solar Inc	15 MW	15 MW	125 MW	125 MW
EverQ	30 MW	30 MW	100 MW	180 MW
Fluitecnik			20 MW	40 MW
G24 Innovation		30 MW	30 MW	200 MW
GE Energy	30 MW	30 MW	30 MW	30 MW
Gintech	60 MW	90 MW	230 MW	560 MW
Helios Technology srl	8 MW	10 MW	10 MW	10 MW
Hellas Solar Cells			30 MW	60 MW
Isofoton	90 MW	90 MW	135 MW	190 MW
JA Solar	75 MW	175 MW	420 MW	585 MW
Jetion Science	25 MW	25 MW	25 MW	25 MW
Jinglong Group	175 MW	175 MW	180 MW	180 MW
Kyocera Corporation	172 MW	172 MW	220 MW	240 MW
Microsol International FZE	14 MW	45 MW	45 MW	45 MW
Millenium Electric	6 MW	6 MW	20 MW	20 MW
Mitsubishi Electric Corporation	135 MW	135	135 MW	135 MW
Mosel Vitelic		12 MW	45 MW	60 MW
Moser Baer		40 MW	80 MW	80 MW
Motech Industries Inc	120 MW	180 MW	300 MW	580 MW
Neo Solar Power		60 MW	90 MW	90 MW
Ningbo Shanshan Ulica		30 MW	80 MW	100 MW
Ningbo Solar Energy Power Co	50 MW	50 MW	200 MW	200 MW
Perfect Energy			20 MW	200 MW
Photon Semiconductor & Energy Co., Ltd.	10 MW	10 MW	40 MW	40 MW
Photovoltech NV SA	30 MW	60 MW	80 MW	80 MW
Photowatt International SA	40 MW	60 MW	100 MW	100 MW
Q-Cells	420 MW	540 MW	590 MW	900 MW
REC Scancell	45 MW	45 MW	145 MW	145 MW
Sanyo	165 MW	260 MW	355 MW	355 MW

TABLE 5.8

(Continued)

Cell (Module) Company	2005	2006	2007	2008
Scheuten Solar	5 MW	18 MW	20 MW	35 MW
Schott Solar	130 MW	130 MW	140 MW	190 MW
Shanghai Chaori		100 MW	150 MW	150 MW
Shanghai Solar Energy S&T		25 MW	30 MW	35 MW
Shanghai TopSola Green Energy Co.	15 MW	15 MW	15 MW	15 MW
Shanshan Ulica Solar	15 MW	15 MW	15 MW	15 MW
Sharp	500 MW	710 MW	850 MW	850 MW
SMIC	3 MW	3 MW	3 MW	25 MW
Solar Cells Hellas		10 MW	60 MW	60 MW
Solar Energy			5 MW	10 MW
Solar Fabrik		50 MW	50 MW	50 MW
Solar Power Industries	35 MW	35 MW	35 MW	35 MW
Solarfun	25 MW	80 MW	240 MW	360 MW
Solarwatt	8 MW	10 MW	15 MW	25 MW
SolarWorld	200 MW	240 MW	450 MW	800 MW
Solland Solar	20 MW	60 MW	110 MW	170 MW
Solsonica			15 MW	30 MW
Spheral Solar		20 MW	20 MW	20 MW
Suniva			10 MW	25 MW
SunPower Corporation	63 MW	100 MW	100 MW	250 MW
Suntech Power Co Ltd	300 MW	300 MW	540 MW	1000 MW
Sunways AG	25 MW	30 MW	45 MW	100 MW
Suzhou Shenglong		12 MW	24 MW	24 MW
Tianda photovoltaic	35 MW	100 MW	100 MW	100 MW
Topray Solar	50 MW	50 MW	50 MW	50 MW
Trina Solar Energy	15 MW	150 MW	150 MW	365 MW
Wuhan linuo Solar Energy	6 MW	6 MW	60 MW	60 MW
Wuxi Solar	30 MW	30 MW	30 MW	30 MW
X Group		2 MW	15 MW	50 Mw
XL Telecom			60 MW	60 MW
Yingli Solar	60 MW	100 MW	200 MW	200 MW
Yunnan Tianda		10 MW	30 MW	60 MW
Zhejiang Global Solar Energy		40 MW	100 MW	140 MW

The current scenario is that a large number of solar PV companies continue to expand their manufacturing capacity, and big capacity solar PV projects are expected in 2009. The expanded fab will provide higher economies of scale along with a reduction in capital outlay for the wafer-based and thin film PV manufacturers. Many companies have even secured long-term supply agreements in order to capitalize on these fab expansions. For example, Suntech Power Holdings Co. has signed a supply agreement with GCL Silicon Technology Holdings Inc. Under this agreement GCL Silicon will supply Suntech with 9,420 metric tons of polysilicon and 1.1 GW of silicon wafers from 2008 to 2012. Q-Cells has signed the BSI/Timminco contract for UMG-Si for a supply of 410 tons in 2008, 3,000 tons in 2009, and 6,000 tons per annum from 2010 until 2013.

MEMC Electronic Materials Inc. entered into a $3.5 billion supply deal with Tainergy Tech Co. Ltd., Taiwan. MEMC would supply solar wafers for a period of 10 years under the terms of the deal. LDK Solar signed a 10-year supply contract with Canadian Solar for the supply of 800 MW of solar PV modules. Trina Solar Ltd. entered into a three-year sales agreement with GreenergyCapital, under which the company would supply GreenergyCapital with PV modules at a total value of $158 million.

Geographically, Germany, Spain, the United States, and China are most active in expanding the production capacities for modules. In the case of Germany, Spain, and United States, the investments are being driven by local demand and the need to reduce operational costs. One way they are trying to do so is by manufacturing the modules close to the installation areas, capitalizing on the location to reduce operational costs.

China already has one of the highest capacities for module production and is following a different strategy in fab expansions altogether. Chinese manufacturers are taking advantage of low labor costs in order to improve their competitiveness and expand their fabs.

Japan currently has the second largest module production capacity in the world because of the strong incentive schemes it established early in this decade, but Japan did not increase its capacity much from 2007 to 2008.

CHAPTER 6

Pulling the Trigger

The risk and volatility factors discussed in Chapter 5 give investors a grasp of the fundamental forces that drive solar stock prices, but of course smart investment decisions take into account much more than fundamentals. Knowing when to buy or sell is the critical component of successful investing.

It's impossible to time the market, but it is possible to design a carefully planned risk/reward calculus—one that uses both quantitative and qualitative analysis—to guide decisions about when to buy or sell. This chapter defines what both long-term and short-term investors and traders must consider about getting in too soon or too late and about assessing if they know enough to pull the trigger.

RISK-REWARD CALCULUS
FOR SOLAR PV STOCKS

Key information that an investor would want to know about solar stocks in his or her portfolio is: What is the risk-reward calculus of the stock? The first step in answering this question is to analyze how the stock moves in relation to the broader market. "Correlation" and "beta" are statistical measures that help us capture how stocks move relative to broad market indices. These measures quantify the risks associated with swings in stock price and help investors arrive at an appropriate reward for taking on this risk.

We consider 28 solar stocks for analysis of the correlation and beta with four indices: the Dow Jones Wilshire 5000 Composite index (DWC), Dow Jones Industrial Average (INDU), Nasdaq Composite index (CCMP), and S&P 500 (SPX). The reason we have picked these 28 stocks is that they offer a good representation of companies across different positions in the supply chain, geographies, and also the different functions (photovoltaics, manufacturing, PV equipment manufacturing, etc.) in the industry.

Beta can be estimated for individual stocks against different stock market indices. In this case, we consider the DWC, INDU, CCMP, and SPX indices, with particular attention paid to the DWC index. The reason for singling out this index is that it is a market capitalization-weighted index of all stocks actively traded in the United States.

In the table that follows, the highlighted column (which represents the correlation and beta values) describes how each solar stock moves with the DWC index. In case a stock has a correlation value of zero, it means that its price is not at all correlated with the DWC index. A positive correlation means that the particular solar stock generally follows the DWC index. A negative correlation implies that the solar stock inversely follows the DWC index; the stock price generally decreases if the DWC index moves up.

The correlation value indicates how likely it is that a particular stock will rise or fall along with the index, but it doesn't describe the degree to which that stock will rise or fall. For that information we need the beta value. The beta value shows how volatile a stock is in relation to the DWC index. A beta value of less than 1 means that the stock will be less volatile than the index; a beta value greater than 1 means that the stock will be more volatile than the index. The higher the beta value, the more volatile the stock. As with most technology stocks, solar stocks generally have high beta values, as Table 6.1 indicates. Table 6.1 provides a ranking of companies with greater than $300 million market capitalization on the basis of the correlation and beta values with respect to the DWC index.

TABLE 6.1

Correlation and Beta Values

CORRELATION	DWC Index	INDU Index	CCMP Index	SPX Index	BETA	DWC Index	INDU Index	CCMP Index	SPX Index
AMAT US Equity	0.56	0.54	0.62	0.56	SOL US Equity	3.61	3.18	2.97	3.38
WFR US Equity	0.54	0.49	0.57	0.51	YGE US Equity	2.97	2.84	2.57	2.86
ESLR US Equity	0.43	0.37	0.44	0.40	TSL US Equity	2.61	2.44	2.18	2.52
SPWR US Equity	0.42	0.37	0.44	0.40	JASO US Equity	2.47	2.29	1.99	2.37
ENER US Equity	0.41	0.34	0.41	0.38	SOLF US Equity	2.06	1.85	1.68	1.94
EMKR US Equity	0.41	0.36	0.44	0.38	FSLR US Equity	2.03	1.95	1.66	1.97
TSL US Equity	0.39	0.34	0.40	0.37	CSIQ US Equity	1.96	1.82	1.67	1.84
JASO US Equity	0.38	0.33	0.37	0.36	WFR US Equity	1.92	1.85	1.74	1.84
YGE US Equity	0.38	0.34	0.40	0.36	EMKR US Equity	1.88	1.80	1.75	1.77
AIXG US Equity	0.37	0.35	0.36	0.36	ESLR US Equity	1.81	1.67	1.61	1.73
STP US Equity	0.35	0.31	0.36	0.33	CSUN US Equity	1.70	1.50	1.52	1.57
FSLR US Equity	0.35	0.31	0.34	0.33	ENER US Equity	1.70	1.51	1.45	1.60
SOL US Equity	0.34	0.28	0.33	0.32	SPWR US Equity	1.69	1.57	1.51	1.59
BTUI US Equity	0.31	0.28	0.33	0.29	LDK US Equity	1.69	1.57	1.51	1.59
CSIQ US Equity	0.29	0.25	0.30	0.27	WWAT US Equity	1.69	1.57	1.39	1.63
SOLF US Equity	0.26	0.22	0.26	0.24	STP US Equity	1.48	1.39	1.31	1.40
WWAT US Equity	0.26	0.22	0.25	0.25	AIXG US Equity	1.31	1.33	1.10	1.29
CSUN US Equity	0.18	0.15	0.19	0.16	BTUI US Equity	1.26	1.20	1.14	1.18
LDK US Equity	0.17	0.14	0.16	0.16	AMAT US Equity	1.10	1.15	1.05	1.10
AKNS US Equity	0.11	0.09	0.09	0.10	AKNS US Equity	0.85	0.73	0.56	0.77
SOEN US Equity	0.11	0.08	0.10	0.09	SOEN US Equity	0.66	0.54	0.51	0.58

As can be seen, the correlation values for all the stocks in the highlighted column are positive, and hence these are all positively correlated to the DWC index. Also, all the stocks are positively correlated to the INDU, CCMP, and SPX indices.

Ranking the stocks on the basis of their respective correlation values with the DWC index (Figure 6.1) reveals that the top five stocks (those showing the maximum correlation with the index) are: Applied Materials Inc. (AMAT), MEMC Electronic Materials (WFR), Evergreen Solar Inc. (ESLR), Sunpower Corp (SPWR), and Energy Conversion Devices (ENER).

The bottom five stocks (those least correlated to the DWC index) are shown in Figure 6.2. These stocks include Solar Enertech Corp (SOEN), Akeena Solar (AKNS), LDK Solar Co. Ltd. (LDK), China Sunergy Co. Ltd. (CSUN), and Worldwater & Solar Technologies (WWAT).

When the stocks are ranked according to the respective beta values with the DWC index (Figure 6.3), Renesola Ltd. (SOL) has the highest beta value of 3.61; followed by Yingli Green Energy (YGE), with the beta value of 2.97; and Trina Solar Ltd. (TSL), with the beta value of 2.61. JA Solar Holdings Co. Ltd. (JASO) and Solarfun Power Hold (SOLF) figure at the fourth and fifth positions in the ranking, with respective beta values of 2.47 and 2.06.

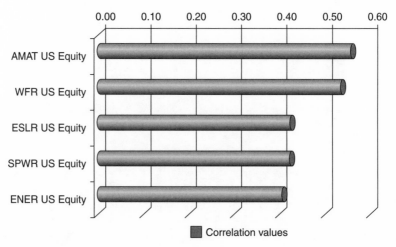

Figure 6.1 Stocks most correlated with DWC index

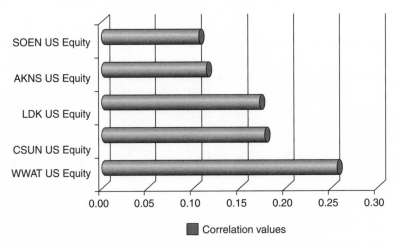

Figure 6.2 Stocks least correlated with the DWC index

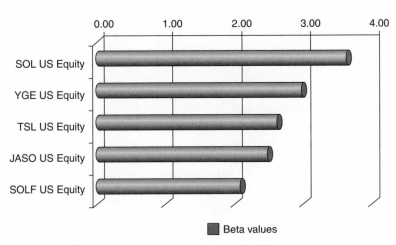

Figure 6.3 Stocks with highest beta with the DWC index

Similarly, we can also study the stocks that have the least beta values with the DWC index (Figure 6.4); any movement in the index would usually be accompanied by the least movement in these stocks. Among the set of stocks analyzed, Solar Enertech Corp. (SOEN) and Akeena Solar Inc. (AKNS) have the least beta, of 0.66 and 0.85, respectively. The third, fourth, and fifth least

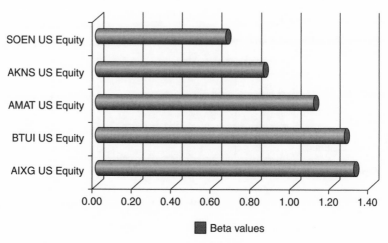

Figure 6.4 Stocks with least beta with the DWC index

beta stocks are Applied Materials Inc. (AMAT), beta 1.1; BTU International Inc. (BTUI), beta 1.26; and Aixtron AG (AIXG), 1.31. The maximum beta is 3.61 and the average beta is 1.83.

CORRELATION AND BETA OF SOLAR PV STOCKS WITH OIL AND NATURAL GAS INDICES

The same 28 solar PV stocks can also be analyzed for correlation and beta with two indices that measure adjacent markets in the broader energy sector: the West Texas Intermediate (USCRWTIM index) and Henry Hub Natural Gas (NGUSHHUB index). Analyzing solar stocks against these indices projects how solar compares against the overall energy sector.

West Texas Intermediate Index

We chose West Texas Intermediate for the oil comparison because it is used as a benchmark in oil pricing and is the underlying commodity of New York Mercantile Exchange's oil futures contracts. West Texas Intermediate is light crude that contains about 0.24 percent sulfur. Its properties and production site make it ideal for being refined in the United States. It is mostly refined in the Midwest and Gulf Coast regions.

Henry Hub Index

The Henry Hub index was chosen for comparison with natural gas because it is the pricing point for natural gas futures contracts traded on the New York Mercantile Exchange (NYMEX). Henry Hub is actually a point on the natural gas pipeline system in Louisiana that is owned by Sabine Pipe Line LLC. The Henry Hub interconnects with nine interstate and four intrastate pipelines, namely: Acadian, Columbia, Gulf Transmission, Gulf South Pipeline, Bridgeline, NGPL, Sea Robin, Southern Natural Pipeline, Texas Gas Transmission, Transcontinental Pipeline, Trunkline Pipeline, Jefferson Island, and Sabine.

Henry Hub is an appropriate choice for comparison between the natural gas and solar sector stock movements because the spot and future prices at Henry Hub, denominated in $/mmbtu (millions of British thermal units), are generally the primary price set for the North American natural gas market. Also, the North American unregulated wellhead and burner-tip natural gas prices are closely correlated to those set at Henry Hub.

Comparison of Solar with Oil

Ranking on the basis of the correlation values with the West Texas Intermediate index is as follows:

Rank 1: JA Solar Holdings Co Ltd, (correlation value of 0.08)

Rank 2: Renesola Ltd (0.05)

Rank 3: Energy Conversion Devices (0.03)

Rank 4: Aixtron AG (0.03)

Rank 5: Emcore Corp (0.03)

Rank 6: First Solar Inc (0.02)

Rank 7: MEMC Electronic Materials (0.02)

Rank 8: SunPower (approximately zero)

Rank 9: Solar Enertech Corp (approximately zero)

Rank 10: Suntech Power Holdings (approximately zero)

Ranking on the basis of the beta values with the West Texas Intermediate index is:

Rank 1: Renesola Ltd, (beta value of 0.23)

Rank 2: JA Solar Holdings Co Ltd (0.21)

Rank 3: Energy Conversion Devices (0.07)

Rank 4: Emcore Corp (0.06)

Rank 5: First Solar Inc (0.05)

Rank 6: Aixtron AG (0.05)

Rank 7: MEMC Electronic Materials (0.03)

Rank 8: SunPower (0.01)

Rank 9: LDK Solar Co. Ltd. (0.01)

Rank 10: Solar Enertech Corp (approximately zero)

The maximum beta is 0.23, and the average beta is –0.002.

Comparison of Solar with Natural Gas

Ranking on the basis of the correlation values with the Henry Hub natural gas index is:

Rank 1: China Sunergy (correlation value of 0.10)

Rank 2: Worldwater and Solar Technology (0.07)

Rank 3: MEMC Electronic Materials (0.06)

Rank 4: Canadian Solar (0.06)

Rank 5: Trina Solar Limited (0.06)

Rank 6: BTU International (0.06)

Rank 7: Akeena Solar Inc. (0.05)

Rank 8: Yingli Green Energy (0.04)

Rank 9: Suntech Power Holdings (0.03)

Rank 10: Emcore (0.02)

Ranking on the basis of the beta values with the Henry Hub natural gas index is:

Rank 1: China Sunergy (beta value of 0.13)

Rank 2: Worldwater and Solar Technology (0.11)

Rank 3: Akeena Solar Inc. (0.08)

Rank 4: Canadian Solar (0.07)

Rank 5: Trina Solar Limited (0.06)

Rank 6: MEMC Electronic Materials (0.05)

Rank 7: BTU International (0.05)

Rank 8: Yingli Green Energy (0.05)

Rank 9: Suntech Power Holding (0.03)

Rank 10: Emcore (0.03)

The maximum beta is 0.13, and the average beta is 0.03.

While oil is not the major source of electricity generation in the United States, the correlation analysis is still useful because it helps us see how solar stock prices move with energy prices.

Clearly, correlation values are very low, meaning that solar stock prices largely do not track with energy prices on a daily basis. Indeed, we can argue that we would expect this to be the case because a lot of other factors would have a more significant bearing on solar stock prices on a daily basis.

However, we would expect solar stock prices in the long run to reflect energy/electricity prices because, at the end of the day, solar PV is a way to generate electricity. Since the industry is relatively nascent, sufficient long-term data to actually compare how long-term, fundamental shifts in oil prices have impacted solar stocks are not available. As the industry matures and access to long-term trends in energy/electricity prices vis-à-vis the prices of solar stocks becomes available, we will be able to derive better insights.

How might these relations appear? For example, we might find that when energy/electricity prices have stayed high over a long-term period, solar stocks have remained at elevated levels as well (for example, in terms of P/E or P/S relative to normal levels). We may also find that when energy/electricity prices have remained depressed over a period of time, solar stock prices have also been depressed over that period in terms of their P/E, P/S, and so on. In general, such a relationship may hold true for other sources of alternative energy as well. Again, drawing these insights requires longer-term data where we can study the relationships between solar stock prices and energy/electricity prices in greater detail.

In this sense, investors' perceptions of where energy and electricity prices are headed during their investment horizon are a crucial input into their investment decision. An investor who believes that energy prices will continue to stay at elevated levels (or increase to even higher levels) would generally be prepared to pay higher multiples for solar stocks (indeed, not just for solar stocks, but also for other types of alternative energy stocks).

COEFFICIENT OF VARIATION OF SOLAR STOCKS AND STANDARD DEVIATION OF THE INDICES

Coefficient of variation is calculated by dividing the standard deviation value by the mean value (mean of the stock values, in our case). The coefficient of variation is a more useful analytic tool than the standard deviation, because the latter can be analyzed effectively only in the context of the mean of the data. Comparing the standard deviation of solar stocks with that of the S&P 500 is apples and oranges. The coefficient of variation is not limited by that; it can be used to compare different data sets with more accuracy. In our case also, if we take a look at the standard deviation values of the solar stocks and the indices, we see that there is a lot of variation, which further highlights the fact that coefficient of variation is a far more efficient means for our analysis.

Standard deviation is a measure of the dispersion of a collection of values, in this case the dispersion of the solar stock prices compared to the four stock indices: the Dow Jones Wilshire 5000 Composite Index (DWC), Dow Jones Industrial Average (INDU), Nasdaq Composite index (CCMP), and S&P 500 (SPX). Again, we'll use the same 28 solar PV stocks as in the examples above.

Standard deviation is used to calculate the historical volatility of the stock. A large standard deviation indicates that the data points (that is, stock prices) are far from the mean, and a small standard deviation indicates that they are clustered closely around the mean. A volatile stock will have a high standard deviation, while the deviation of a stable stock will be lower.

The standard deviation for each of the 28 solar stocks (dates of calculation are either the last two years or the IPO date, whichever is more recent) are as follows:

Applied Materials Inc.: 1.5
MEMC Electronic Materials: 16.5

First Solar: 92
Suntech Power Holdings: 12.5
SunPower Corporation: 28
LDK Solar Holdings: 10
JA Solar Holdings Co. Ltd.: 6
Yingli Green Energy Holding: 7
Aixtron: 4
Evergreen Solar Inc.: 2
Energy Conversion Devices: 11
Trina Solar Ltd.: 12
Solarfun Power Holding: 5
Renesola Limited: 5
Emcore: 2.5
Canadian Solar: 10
Worldwater and Solar Technology: 0.6
China Sunergy Corporation Ltd.: 2.5
Ascent Solar Technologies: 6
Akeena Solar Inc.: 2
Spire Corporation: 4
Amtech Systems Inc.: 2.6
Daystar Technologies Inc.: 3
BTU International Inc.: 2.5
Solar Enertech Corp.: 0.4
XSUNX Inc.: 0.5
Open Energy Corporation: 0.5
GT Solar: 1.6

The standard deviations of the respective indices are:

Dow Jones: 929
Nasdaq Composite: 177
Wilshire 5000: 915
S&P 500: 90

As the huge variations in the values from the data above indicate, we cannot analyze using only the standard deviation values.

Moving on to the coefficient of variation, the respective value for all solar stocks (dates of calculation are either the last two years or the IPO date, whichever is more recent) is:

Applied Materials Inc.: 0.8
MEMC Electronic Materials: 0.3
First Solar: 0.6
Suntech Power Holdings: 0.3
SunPower Corporation: 0.5
LDK Solar Holdings: 0.3
J A Solar Holdings Co. Ltd.: 0.4
Yingli Green Energy Holding: 0.3
Aixtron: 0.5
Evergreen Solar Inc.: 0.2
Energy Conversion Devices: 0.3
Trina Solar Ltd.: 0.3
Solarfun Power Holding: 0.35
Renesola Ltd.: 0.3
Emcore: 0.3
Canadian Solar: 0.6
Worldwater and Solar Technology: 0.7
China Sunergy Corporation Ltd.: 0.25
Ascent Solar Technologies: 0.6
Akeena Solar Inc.: 0.5
Spire Corporation: 0.4
Amtech Systems Inc.: 0.3
Daystar Technologies Inc.: 0.5
BTU International Inc.: 0.2
Solar Enertech Corp.: 0.3
XsunX Inc.: 0.7
Open Energy Corporation: 0.7
GT Solar: 0.1

Coefficients of variation of the respective indices are:

Dow Jones: 0.08
Nasdaq Composite: 0.08

Wilshire 5000: 0.07

S&P 500: 0.07

In order to compare the solar stock coefficient of variation with the four index values, we calculate the average solar stock coefficient of variation: 0.4.

We observe that the average coefficient of variation of solar stocks (0.4) is much higher when compared to the coefficients of variations of Wilshire (0.07), S&P 500 index (0.07), Dow Jones (0.08), and Nasdaq Composite (0.08).

This is also expected, as the volatility within the solar stocks will be higher when compared to the volatility in the respective indices.

Quantitative analysis of the kind discussed above is a powerful investment tool, but the potential upside and downside of a given solar stock can't be expressed by numbers alone. Qualitative factors also need to be considered, and it is to those that we now turn.

DIFFERENTIATION FOR THE LONG-TERM INVESTOR

One may argue that over time the solar PV industry will evolve into a commodity industry where the only thing that matters is cost per watt. However, things are not that simple. There are several other parameters that could potentially differentiate one technology from another.

The U.S. Department of Energy (DOE) uses the example of the bicycle industry to highlight how differentiation might occur ("Venture Investing in Solar, 2007/2008: Investment Theses, Market Gaps and Opportunities," presentation by Craig Cornelius, program manager):

- Aluminum bicycles have the advantage of manufacturability.
- Steel bicycles have the lowest prices.
- Titanium bicycles are the most durable.
- Carbon-based bicycles give the best performance.

This example highlights the important point that even as an industry becomes increasingly commoditized, there may be areas where manufacturers may be able to distinguish their products. What could these areas be in the case of solar PV? In other words, in what areas could solar PV manufacturers claim that their product is unique?

Lowest Costs per Watt

The first and perhaps most obvious differentiating feature for a company is its ability to achieve lowest costs per watt, translating into low and consistent (level) electricity costs. While the advantages of lowest costs per watt are clear, there is a hidden cost of space/area that is not captured in this number. Indeed, for space-constrained installations, one may find it more economical to have a higher power output per unit area than to just go with least-cost-per-watt panels. The current leader in costs per watt is First Solar; indeed, this company also seems the one best poised to reach grid parity first.

Best Fit for Specific Applications

Some solar PV products may be specially well-suited to certain applications. For example, for building integrated applications, lightweight, flexible panels may be the best choice because they are easier to install and also because it may be possible to achieve greater roof coverage due to their lightweight nature. The commercial rooftops segment is one area where this is proving to be a major differentiator. Energy Conversion Devices produces its Uni-Solar laminates, which are especially well-suited to building-integrated PV applications.

High Power Output/Best Efficiency

In space-constrained applications, conversion efficiency may well be the primary criterion to determine which solar panel to use. SunPower, with its industry-leading efficiencies, has a differentiated product in this context. Besides having the highest efficiency commercially available panels, SunPower also has attractive aesthetics. This is another differentiating factor in the solar PV industry.

Lowest Silicon Usage

Some crystalline-silicon-based solar PV companies may have lower silicon consumption per watt than others, and this could prove to be a differentiating factor. For example, Evergreen Solar, with its proprietary string ribbon technology, has the lowest silicon usage, as compared to other crystalline silicon companies, and is poised to bring this down even further. This could differentiate Evergreen Solar (and other companies that might have low silicon usage) from competitors.

The U.S. Department of Energy (DOE) describes criteria for success in the solar PV industry for both crystalline silicon and thin film companies.[1] Here are some of these.

Success in crystalline silicon PV:

- Ability to integrate vertically or to license products
- Ability to ramp up and achieve scale
- Ability to survive a potential industry shake-out in 2010 driven by companies that have achieved scale

Success in thin film PV:

- Ability to ramp up (a key factor here is the level of complexity in the deposition process)
- Ability to manage with potentially lower margins
- Achieving and demonstrating product reliability

TRIGGER FOR THE LONG-TERM INVESTOR

These qualitative factors are important for both short- and long-term investors, but the latter should be particularly focused on them. An investor with a reasonably long time horizon would be primarily concerned about how the company is positioned competitively and whether the fundamentals of the company justify a long-term investment. More specifically, a long-term investor would be deeply concerned about the following aspects of a company before deciding to invest in it:

- Potential for sales and earnings growth
- How the company is placed vis-à-vis competition, and how this could possibly change in the future

- Potential for the company to consolidate and expand its position in the market
- Ability of the company to maintain and extend its technology advantages

FIRST SOLAR FROM A LONG-TERM PERSPECTIVE

By all these measures, a company like First Solar fits the bill for an investor with a long time horizon. The company has posted excellent revenue and earnings growth in the past and is poised to continue on this track for some time to come. Revenue in Q2 2008 was ~$267 million, compared to ~$77 million in Q2 2007, a year-on-year increase of 247 percent. Production in Q2 2008 was 114.1 MW. With large capacity ramps planned and a strong track record in execution, First Solar is well-positioned on the production side. Indeed, once all lines in Malaysia ramp up (by H2 2009), First Solar's aggregate (on a global basis) manufacturing capacity would be greater than one GW.

First Solar has also enjoyed exceptional gross margins of 50-plus percent. Gross margins in Q2 2008 came in at 54.2 percent, and in Q1 2008 at 53 percent. This in turn is made possible by First Solar's industry-leading control over costs per watt: in Q2 2008 it was just $1.18 (in fact, this number was pushed higher by six cents because of ramp-related costs and two cents due to foreign exchange related impact). With improvements in manufacturing technology and processes, this number is poised to go down further, offsetting at least part of expected declines in ASPs. This leaves the company at a significant advantage compared to other companies in the solar PV industry.

Long-Term Sales Contracts

At the end of 2007, FSLR had sales contracts with 12 customers in the European Union (EU) to the tune of ~ 4.5 billion. These would cover aggregate sales of 3.2 gigawatts of solar PV modules from 2008 through 2012. First Solar's customer list includes companies such as Conergy AG, Juwi Solar GmbH, and Phoenix Solar AG.

We note that these long-term contracts contain clauses specifying an annual increase in watts per module and a decrease in ASPs. First Solar's cost reduction plans need to progress at least at these rates in order that margins be maintained.

These large-scale, long-term sales contracts mean First Solar has good revenue visibility several years into the future. This also means that First Solar occupies a leading position in terms of sales contracts.

First Solar has traditionally been very dependent on Germany for its revenues, but over time it looks poised to expand into other fast-growing markets as well.

Leading in Technology: Cadmium Telluride Thin Film

First Solar's modules are based on single junction thin film with cadmium telluride as the absorption layer and cadmium sulfide as the window layer. The semiconductor material used by First Solar is well matched to the solar spectrum, so this allows First Solar modules to deliver conversion efficiencies of ~10.7 percent (as of Q2 2008).

Under low light and/or high temperatures, First Solar's modules perform better than traditional silicon-based cells. This means the kilowatt hours generated per rated watt is higher than for a comparable PV module based on silicon.

Some of the advantages of First Solar's technology:

- Accomplishes conversion efficiencies of ~10.7 percent using just 1 percent of the semiconductor material used by traditional silicon-based modules.
- Has integrated manufacturing, with proprietary technologies (for instance, vapor transport deposition technology for thin film coating) that cannot be easily replicated by competitors.
- The steps of semiconductor deposition, cell definition, and assembly/testing are accomplished in a matter of just 2½ hours, as against the time-consuming, discrete steps of manufacturing crystalline silicon-based PV.
- Company has a "Copy Smart" methodology that allows it to replicate lines faster.

Emerging Business Models

First Solar has traditionally targeted solar PV applications requiring ground-mounted installations (usually >one MW) and also commercial roof-mounted systems (30 KW to one-plus MW). Indeed, targeting this market has helped First Solar become a leading company in the solar PV industry.

However, with new business models potentially emerging, First Solar is developing relationships with U.S. utilities to address this market and is looking to gain a share of the utility market in states that have Renewable Portfolio Standards.

In an analyst conference call, Michael Ahearn, chairman and CEO, described different types of pilot projects to address three different business models that might emerge in a big way some time in the future:

- IPP model—to build and operate solar power plants and sell system output to regulated utilities
- Building a turnkey system and transfer ownership to regulated utilities
- Building a turnkey system and selling it to unregulated affiliates of energy companies

These pilots allow First Solar to capture value in these markets as and when they emerge. Also, given its cost advantages compared to other companies, these business models might be particularly attractive for FSLR. That said, emerging business models may affect valuation either positively or negatively. Investors must evaluate changing business models in terms of changes to valuation. At this juncture, there is simply insufficient information to judge how these changes will affect stock prices.

Path to Grid Parity

As mentioned earlier, First Solar already has the lowest costs per watt in the solar PV industry. Moreover, the company has a clear plan of action in order to reach grid parity in the coming years. That path is outlined below ("First Solar Introduction," presentation by First Solar, April 2008, accessed on September 10, 2008

at http://www.texassolarforum.com/downloads/Gruber_TSF_
Austin_42508.pdf):

1. The company is targeting grid parity by 2010 to 2012.
2. For comparison purposes, consumer costs of electricity are
   assumed to be 8 to 10 cents per kilowatt hour. This is the
   point at which solar reaches grid parity and can compete
   on costs with other sources of electricity.
3. This would mean First Solar would have to reach installed
   system costs of $2.00 to $2.50 per watt.
4. In terms of module prices, this translates to $1.00 to $1.25
   and a corresponding BOS reduction.

Once grid parity is reached, the market for First Solar might
potentially increase dramatically beyond its current market, since
it could become the first player in the industry to achieve this.
Becoming cost competitive with coal and other traditional ways of
generating electricity can open up several opportunities for First
Solar, which could potentially benefit an investor with a sufficiently
large investment horizon.

As we have described, First Solar presents potential benefits on
several fronts for the long-term investor. For an investor wanting to
participate in the rapid growth of the solar industry, therefore, First
Solar would appear to be one of the front-runners. If a long-term
investor indeed decides to select First Solar stock for inclusion in
her portfolio, a good strategy might be to follow a cost-averaging
strategy where she accumulates the stock at different price levels
over time, in the expectation of a reasonable average buying price.

Such an investor would not be particularly bothered about
regular ups and downs in the market price of the stock; rather, she
will invest in the belief that with strong fundamentals, superior
competitive positioning, and a clear road map to attain grid parity,
First Solar could give good returns on investment.

SUNTECH POWER FROM A
LONG-TERM PERSPECTIVE

As described at the beginning of this chapter, we're not concerned
about timing, but rather, about confidence in stocks and knowing when to take action. In this context, Suntech Power is another

quality solar PV company with strong fundamentals and that has proved, time and again, its ability to execute.

In terms of its financials, Suntech Power has posted strong revenue growth of 50-plus percent in Q2 2008 on a year-on-year basis. At the end of Q2 2008, production capacity was 660 MW, which immediately indicates the ability of the company to expand capacity on a large scale. Indeed, as and when the industry moves toward consolidation, Suntech Power may be attractively positioned to capture value. That said, Suntech's raw material strategy, like that of many others, may have been fundamentally flawed during the boom time when the company made a number of take-or-pay contracts and invested heavily upstream in polysilicon ventures that may be permanently underwater should the downturn in solar persist.

The company has a robust set of contracts to manage its raw material requirements, which is especially important given its capacity expansion plans. To recapitulate, Suntech Power's three-pronged strategy for raw materials supply management intended the following:

- Long-term supply contracts at competitive prices, while at the same time maintaining some degree of price and volume flexibility.
- Building financial strength by raising capital, so the company will be able to finance not only its own expansion, but the expansion of its silicon suppliers. This is particularly important in the solar PV industry, since many polysilicon contracts require significant prepayments.
- Streamlining the supply chain by developing facilities (at the Wuxi Solar Park) that will put suppliers in close proximity to its own facilities.

Suntech Power also has a strong research and development setup, which ensures that the company remains at the cutting edge of solar PV technology. The company's R&D has five main thrust areas (Suntech Power, annual report 2007, 10-11):

- Pluto technology, which seeks to increase conversion efficiency by 10-plus percent while maintaining costs of production.

- Research in manufacturing process-related innovations leading to a greater degree of automation in operations.
- New product development; for example, the development of solar modules, in collaboration with Akeena Solar, that have superior features, including a better aesthetic appearance.
- Materials research: Suntech Power seeks to make improvements in materials (silicon feedstock, back sheets) that are used in solar PV manufacturing.
- Thin-film-based PV products: besides crystalline silicon-based PV, Suntech Power also researches thin film PV.

Suntech Power is also attractively positioned in the building integrated photovoltaics market with its "MSK Solar Design Line" products.

Given the fundamentals of Suntech Power, a typical investor looking to invest in the solar sector would probably want to know how the company's supply chain strategy and cash flow needs will unfold before investing in the company.

KNOWING WHEN TO TAKE ACTION

An investor may believe that a company has excellent prospects for revenue and earnings growth and a strong management in place. However, before going ahead and buying the stock, he or she needs to check if it is indeed the best time to buy. Consider the example of an investor contemplating whether to buy stock in a company with a heavy revenue exposure to Spain. As we have seen, Spain is a large market for solar PV, driven by generous subsidies, but these subsidies were drastically curtailed in 2008. On the other hand, an investor who is less risk-averse might want to "play" how the U.S. subsidy picture pans out.

Also, an investor might want to wait until a major peer company or group of companies reports results in order to better gauge the prospects of the industry. These peer companies could even be based in different geographies. For example, management commentary following results at a company like Wacker could yield valuable insights on polysilicon prices in the future. Or earnings and related news flowing from First Solar might give an investor a better idea of how the German PV market is progressing.

To generalize these notions, we can say that an investor would need to weigh the benefits of information that may be available in the near future versus the possible downside in doing so. Indeed, even for a long-term investor, it is worthwhile considering what the best entry and exit point for investing in a stock would be.

There may also be cases where an investor may simply lack adequate information to make an informed decision on a particular stock. For example, when a new technology or methodology (such as upgraded metallurgical silicon) for solar PV manufacturing emerges, it would be difficult for an investor to judge a company that wishes to adopt this technology/methodology. In fact, upgraded metallurgical silicon generated quite a lot of controversy among investors; their concerns about this technology were alleviated (if not fully, at least partially) with the announcement that a reputable company—Q-Cells AG—had found the technology suitable for commercial adoption and with the news of the company's major thrust into upgraded metallurgical silicon-based cell manufacturing.

Recently, several companies in the industry have announced plans for vertical integration. While these might be strategically "correct" decisions, they carry significant execution risks, which are extremely difficult to predict, especially given the increasingly complex nature of manufacturing processes as one moves upstream along the value chain. In such cases, again, an investor might find that there is not enough information to make an investment decision.

As is the case with any other industry, it is important for an investor to know when there is not enough adequate information to invest or to stay invested in a specific stock and make investment decisions accordingly.

TO BUY, SELL, OR SHORT

This section focuses on generalized conditions under which an investor may consider whether to buy, sell, or short a solar stock or a group of solar stocks. Once again, we are not attempting to describe how to time the market, but rather to explain principles an investor might want to use in order to decide when to take action. Let it be clear that these considerations, while they may be helpful, are not substitutes for appropriate actions implied by well conceived valuation measures.

Conditions to Buy

A long-term investor would want to be reasonably sure of the prospects of a company before investing in its stock. Given that several companies in the solar industry have been around for only a decade or so, a long-term investor would want enough information at his or her disposal to be able to make a decision with confidence. Below, we identify a few examples of news flows that could help reassure investors that it might be a good time to buy.

Successful Execution

While news of planned capacity increases may cause a company's stock price to rise in the near term, it gives no indication of how well the company will be able to execute its plans. On the other hand, a long-term investor would feel reassured with news of a company actually executing capacity ramps according to plan.

In some cases, companies may have new technology that requires validation in the eyes of investors. Suntech Power's Pluto technology is an example of this. This technology is meant to improve conversion efficiency (>16 percent on polycrystalline-silicon-based cells, up from 15-plus percent; >18 percent on monocrystalline-silicon-based cells, up from 16-plus percent) without a corresponding increase in costs, which means the company can reduce per watt costs using this technology.

In the solar PV industry, however, it is not always easy to move technology from the lab (or even pilot lines) to actual commercial-scale production. Therefore, investors would want to know whether and when such a technology has been successfully implemented on a large scale. Indeed, any news on successful commercial production would send a positive signal to a long-term investor.

To give another example, when Evergreen Solar first announced plans for its "Quad" furnace, it was not completely evident whether it would be a success. As noted earlier, Quad would allow Evergreen Solar to grow four ribbons of silicon (as compared to two ribbons) and offer several advantages in manufacturing and costs.

As news of successful commercial production at its Devens facility came in, investors had reasons to feel more convinced about the prospects of the company and its stock.

In the case of Pluto for Suntech Power, or Quad for Evergreen Solar, though the technology was new, and investors had at least some degree of reassurance, given Suntech Power's strong track record and also the fact that Pluto is based on crystalline silicon, an area where the company is known to have a significant degree of expertise.

However, when a solar PV company decides to move into a completely new technology, a long-term investor would need a further boost in confidence before deciding to invest. Another example of Suntech Power illustrates this point. Besides Pluto, the company has ambitious plans for thin-film-based PV. This is an area where an investor does not have much in the way of a commercial-scale track record to evaluate the company's execution capabilities, though Suntech Power does have strong R&D efforts (with an array of patents) to back this. Therefore, a long-term investor would need further color on this to be able to arrive at how this foray would impact Suntech Power's valuation.

News Suggesting Reduced Stock Volatility in the Future

We have observed that several solar PV companies have plans for massive capacity expansions and/or backward integration. While this creates opportunities for growth and capture of significant additional value, it could also greatly enhance volatility in the stock because of execution risks. An investor who is otherwise positive about the company has the following choices:

- To buy, with the belief that execution of capacity increases/backward integration happens according to plan
- To wait until there is news that could reduce volatility

For example, an investor who had been positive on Trina Solar except for the risks (and higher volatility) associated with its backward integration plans might have been reassured on hearing in April 2008 that the company had canceled plans to build a $1 billion polysilicon plant due to its belief that poly supplies might ease in the long term. In general, an investor who favors reduced volatility might decide to buy a stock after news events that suggest reduced volatility in the future.

Management Changes
Sometimes, management changes can signal a possible shift in the strategic direction of a company, and if this suggested shift is positive enough and other conditions look favorable, an investor may consider buying the stock.

A good example would be Energy Conversion Devices. Under earlier management the company was heavily focused on R&D and had a history of losses. However, with a new CEO (Mark Morelli) appointed in H2 2007, the company started a transformation with much more focus on the operational aspects of the business. Indeed, in his first analyst conference call after taking over as CEO, Mr. Morelli articulated his vision for the company, backed by clear pointers on how his plans for the Energy Conversion Devices would be achieved operationally. Soon enough, in fact in less than a year, the company did achieve profitability, with its stock price surging in response. Key metrics such as gross margins and costs per watt improved materially during this period.

Although this kind of story is not always easy to predict, it is worth keeping in mind that new management with a clearly articulated strategy (backed by nuts-and-bolts details on how this strategy will be achieved) could signal to an investor that it might be a good time to buy the company's stock.

Conditions to Sell or Short

There are several conditions under which an investor may want to sell or short a solar stock.

Doubts over Technology
The solar PV industry has been traditionally dominated by crystalline-silicon-based players, with thin film (mainly cadmium telluride and amorphous silicon, with CIGS another potentially scalable technology) recently making some inroads. Investors are generally skeptical (at least initially) about any new technology that claims to be the next "new thing," especially given that mass scale, low cost

manufacturing of solar PV is a different proposition altogether as compared to achieving favorable results in the lab.

As First Solar noted in its 10-K filing for FY07, the technology for thin film PV has been in existence for over two decades, but commercial thin-film-based PV had not really taken off until a few years back. As of late 2008, thin-film-based PV accounted for less than 10 percent of the total PV market, though this number is expected to rise significantly in the near future. Therefore, news of production delays or difficulties in ramping can trigger heavy activity in such a company's stock.

Upgraded metallurgical silicon is one such technology where investors raised doubts about its viability. With its claim of low cost silicon of acceptable quality for the solar industry, Timminco's upgraded metallurgical silicon plans have been controversial in the past. Indeed, several investors raised doubts about whether Timminco really had the ability to deliver on its promises. The air of controversy surrounding the company's technology meant that investors were a lot more sensitive and a lot less tolerant of any negative surprises coming from the company.

On April 16, 2008, the company's stock price was at C$28.39. By April 22 the stock price was down nearly 36 percent. to C$18.23, on negative sentiment. On August 12, 2008, the stock price again fell dramatically, by 24 percent to C$15.10, after the company reported a production number that fell short of analyst expectations.

Despite a high-profile agreement with Q-Cells and positive opinions expressed after an independent operational review by *Photon* magazine, the controversy around Timminco does not seem to have evaporated completely.

In general, if an investor has reservations about a company's technology, she may decide to sell the stock or even short it. If an investor does decide on selling or going short, she may want to take the necessary action after the stock has run up, following some positive news (such as a sales contract).

Legal and/or Ethical Concerns

Legal or ethical issues could also prompt an investor to sell or short a solar stock. These may surface either as rumors or as official news releases. On coming to hear of any such rumor or news, an investor would need to decide if it is material enough to necessitate selling or shorting the stock.

Following are examples of such news events that might cause investors to take action:

- News of intellectual property litigation related to a company's core technology/technologies might signal a red flag to investors, prompting them to sell or go short. Indeed, this is an important trigger in the technology-intensive solar PV industry.
- Solar PV companies usually offer warranties for performance of solar modules over an extended period of time (up to 20 to 25 years), and also for product quality for a shorter period (SunPower offers this for up to 10 years). Any news of performance or product issues could therefore send a particularly bad signal to investors, prompting them to sell or short the stock.
- As we note in Chapter 7, some companies could face environmental risks that might jeopardize their financials and stock price. For example, First Solar uses cadmium telluride and cadmium sulfide in its modules, and news of leakage of these hazardous chemicals could trigger adverse reactions on the stock price, sometimes even if these leakages are of an isolated nature.

For a long-term investor the key here again is to discern whether such news events would have a long-lasting impact that would keep stock prices depressed, or whether the swing in the stock price is just temporary.

NOTE

1. "Venture Investing in Solar, 2007/2008: Investment Theses, Market Gaps and Opportunities," presentation by Craig Cornelius, program manager.

CHAPTER 7

Investing for the Long Run

All portfolios are constructed with investment goals in mind. Investors decide the appropriate holding period, risk, and return according to their goals. Therefore, as one reads this chapter, it is vital to understand how much risk your portfolio can handle. Since solar stocks have a level of risk most could not accept as a stand-alone investment, they are perhaps more appropriate as a basket of stocks in a diversified equity portfolio.

As with all portfolios, choosing what to buy and what price to pay must be justified by and relevant to your investment strategy. We can consider the basket of solar stocks independently of its impact on a diversified equity portfolio, but we cannot know the importance and need for investors to consider the basket's impact on their total portfolio. When evaluating the basket, a number of solar stock indices can be used to compare return and risk so the basket's benchmark can be established.

In this chapter we will also discuss the long-term considerations behind sources of risk, rather than the basics of measuring risk using standard deviation and beta. Solar stocks remain shrouded in various risks, but they can be identified, allowing investors to begin the process of stock selection.

IDENTIFYING RISKS

The advantage of choosing a basket of solar stocks for the long-term investor is that it safeguards against overexposure to risk. Currently, it is difficult to predict which specific solar technology will be the winner in the long run. Moreover, the industry will continue to be highly dependent on government subsidies until costs decrease sufficiently to reach grid parity, and it is uncertain how subsidies will change in the future, which makes it difficult to determine the advantages and disadvantages of specific geographic exposures. A portfolio approach serves to minimize the risks of exposure to specific technologies or geographies as well as risks coming from over- or undersupply (of raw materials or modules), slack demand, manufacturing missteps, long warrantees, financing risk, substitution of other renewable energy, macroeconomic factors (like foreign exchange and interest rates), and many other risks discussed in this chapter. Choosing a basket of solar stocks (that is, taking a portfolio approach) can minimize risk and lead to better risk-adjusted returns.

As discussed in previous chapters, two main technologies dominate the photovoltaic market: crystalline silicon (c-Si) and thin film. According to Solarbuzz, a leading source of solar industry information, c-Si accounts for ~93 percent of the market, and thin film accounts for ~7 percent. Apart from these dominant technologies, there are other technologies, such as concentrator photovoltaic, with a minuscule market share currently, but that might prove to be among the eventual winners in the longer term.

Given this situation, it makes sense to invest in a basket of solar stocks representing a variety of technologies. For example, a diversified approach toward technology can be seen in the strategic investments of Moser Baer Photovoltaic, which has invested in crystalline silicon and thin film as well as concentrator photovoltaics. Q-Cells is another company with a portfolio of technologies. It has subsidiary and joint ventures covering cadmium telluride, micromorph, and other thin film technologies (see Table 7.1). While choosing to invest in either Motech in India or Q-Cells in Germany may not be possible for many private investors in the United States, they can serve as examples of how some companies are diversifying their own technology risks in an uncertain future. Investors may benefit

TABLE 7.1

Q-Cells Investment Portfolio

Wafer-Based Technology	Thin Film Business
REC (17.9%)	*Fixed Substrates (Glass)*
• Strategic partner and main supplier	Calyxo (100%)
• Technology leader in polycrystalline	• Cadmium-Telluride technology
silicon production	Brilliant 234 (100%)
	• Micromorph silicon technology
EverQ (33.3%)	Solibro (67.5%)
• String ribbon technology	• CIGS technology
• Joint venture with Evergreen Solar	CSG Solar (21.71%)
and REC	• Crystalline silicon on glass
Solaria (12.39%)	*Flexible Substrates*
• Low-concentration PV technology	VHF-Technologies (23.44% → 51%)
	• Amorphous silicon on plastic foil
	("Bexcell")

Source: Q-Cells AG

from selecting stocks from both groups while more closely examining the technology positions of various companies. Ever-Q, Brilliant 234, and VHF-Technologies have been renamed Sovello, Sontor, and Flexcell, respectively, while this book was in preparation.

Competitive Risks between Different Technologies

The efficiencies of industry average crystalline silicon are higher when compared to thin film. The efficiency of crystalline silicon is ~16 percent, while that of thin film is usually less than 10 percent (with the exception of First Solar, which has ~10.6 percent efficiency). Another major strategic advantage of crystalline silicon is the proven technology with low entry barriers for new entrants. The high efficiency and proven technology of crystalline silicon attract a significant level of capex investments by established players in the industry, which further gives a competitive advantage to the use of this technology.

In the thin film segment, there is competition among the thin film/cadmium telluride on glass and CIGS technology. Both have higher weight due to glass and thus are primarily used for large-scale installations, mostly on commercial rooftops. The competitive advantage with CIGS is that a host of start-up companies are trying to develop a new CIGS technology that might enable CIGS to deliver 12 to 15 percent efficiency. On the other hand, even if possible, this would take around three years, and currently many major CIGS players are moving to traditional crystalline silicon markets because of the encapsulation limitation of CIGS. (When moisture comes in contact with copper, the circuit corrodes; this is known as "encapsulation.")

In the crystalline silicon segment, upgraded metallurgical silicon is a huge bet, and a lot will be determined on whether it pays off. If successful, upgraded metallurgical silicon will significantly reduce the silicon costs, but this technology is still at risk, as it has not been extensively tested on a large scale until now. Q-Cells has planned a major push into upgraded metallurgical silicon as part of its effort to reduce silicon costs. It signed the BSI/Timminco contract for upgraded metallurgical silicon for very high supply volumes of 410 tons in 2008, 3,000 tons in 2009, and 6,000 tons per year from 2010 until 2013. This has exposed Q-Cells to a high level of risk, in case upgraded metallurgical silicon does not result in the expected cell efficiencies or if production yields are not adequate. The risk that the oversupply in polysilicon beginning with the credit crisis persists may also result in the company's write-down of its investments, resulting in a loss of shareholder value.

RESEARCH AND DEVELOPMENT

Given that technology is a key source of competitive advantage in the solar industry, it is important for the investor to know the key technological advantages of each company and how secure these advantages are in terms of intellectual property protection mechanisms.

For example, Evergreen Solar's patented string ribbon technology and its proprietary quad ribbon wafer furnace design may help the company attain better manufacturing efficiencies. It could also possibly lead to an outsourcing or licensing strategy.

First Solar notes in its 10-K filing for FY07 that it has a proprietary vapor transport deposition process and a laser scribing process that are fundamental to its success.

JA Solar, however, has not patented its solar cell manufacturing technology, but instead relies on trade secrets and other mechanisms, protected by confidentiality agreements. Such a system of intellectual property protection might be more susceptible to breaches than a patent-based mechanism. Moreover, as the credit crisis of 2008 resulted in unfavorable macroeconomic factors for the solar industry, investors seemed to lump JA Solar into the group of low-quality undifferentiated companies.

Also, it may be difficult to protect intellectual property rights in certain countries due to the nature of their legal systems. JA Solar, which has most of its operations in and derives most of its revenue from China, notes in an SEC filing, "Enforcement under intellectual property laws in China may be slow and difficult in light of the application of such laws and the uncertainties associated with the [People's Republic of China] legal system."

Given that these are critical issues in a technology intensive industry, it may be useful for the investor to study the intellectual property protection mechanisms of solar companies.

R&D Expenses as a Percentage of Revenue

Since the solar industry is technology intensive, R&D expenses are an important indicator of how well a company is poised in terms of its investments for the future. Even if a company has access to the most advanced technology currently available, it is important to invest in R&D so that this advantage can be sustained. In order to compare R&D expenses on a common basis, consider Table 7.2.

TABLE 7.2

Comparative R&D Expenses by Company

Company	R&D as Percentage of Revenue, 1Q 2008
Uni-Solar/Energy Conversion Devices (ENER)	~3.0%
First Solar (FSLR)	2.4%
SunPower (SPWR)	1.6%
Suntech Power (STP)	0.6%
JA Solar (JASO)	0.2%

GEOGRAPHIC DIVERSIFICATION

When investing in solar stocks, it is also important to account for geographic risks stemming from solar's continued reliance on government subsidy. For example, in 2007 First Solar (FSLR) received greater than 85 percent of its revenues from Germany. An investment in a company whose revenues are heavily concentrated in just a few geographical regions would leave the investor exposed to considerable risk, depending on how the subsidy picture evolves. However, by adopting a portfolio approach with a diversified geographic profile, this risk can be minimized.

SUPPLY VISIBILITY

Solar module manufacturers are dependent on certain key raw material inputs. As discussed in much detail in previous chapters, polysilicon was in short supply for much of 2006 through 2008. The sharp rise in polysilicon prices led to difficulties throughout the value chain as companies with supply visibility gained investor favor, and those without it saw much discounted relative valuations. Therefore, in order to obtain access to polysilicon, several companies struck multiyear deals with polysilicon vendors and were rewarded handsomely with rich valuations as demand seemed "endless," leading investors to pay a premium for the revenue visibility.

However, the benefits to long-term supply agreements have not been fully realized. As 2008 ended, polysilicon prices had begun to fall, and what began as a slowdown created by the credit crisis became an oversupply situation, with falling prices exposing companies without much product differentiation. As this book goes to print, it remains unclear if long-term contracts will be viewed positively or negatively, especially as a wave of new polysilicon supplies result in a growing oversupply.

One example of a company that signed long-term polysilicon supply contracts is Evergreen Solar, which inked multiyear agreements with several companies, including Wacker Chemie AG and DC Chemical. The agreements provide Evergreen with adequate raw material to produce 125 MW in 2009, 300 MW in 2010, 600 MW in 2011, and 850 MW in 2012.

To illustrate the nature of these supply arrangements, Evergreen signed a "fixed price" contract with DC Chemical at the end of January 2008 that runs from 2009 through 2015. As part of this contract, Evergreen had to pay a cumulative nonrefundable amount of ~$36 million by the end of 2008. These contracts ensure that Evergreen has adequate poly supplies to meet its requirements until 2012. Depending on the contract price, Evergreen could be faced with higher polysilicon costs and thereby surrender some of its advantage from its process efficiency should polysilicon prices fall to cyclical lows under $30 per kilogram. On the other hand, if the oversupply of modules is relatively short-lived and polysilicon prices remain at elevated levels, then Evergreen's secure feedstock supplies are likely to once again be viewed as favorable. As solar PV experiences periodic shortages and oversupply, long-term investors must weigh the risk.

As an example of another such contract, SunPower signed a "fixed price" agreement with Hemlock Semiconductor Corporation to obtain poly supplies to support more than two GW of solar cell production from 2010 to 2019. First Solar, which makes thin film PV modules, has long-term supply contracts for the key raw materials tellurium and cadmium telluride. These contracts give the investor good visibility on the supply side.

Suntech Power is a company that, despite having some "fixed term" contracts for polysilicon, would still depend on the spot market "to a significant extent" for supplies in 2008, according to its 20-F filing for FY07.

Trina Solar has an elaborately articulated strategy for managing its raw material supplies. It planned its raw material procurement sufficiently early so it would not be in a disadvantageous position for negotiations later. Indeed, long-term raw material procurement contracts signed in 2007 have helped the company obtain relatively favorable prices. Trina Solar has also tried to keep its supplier base relatively diversified. In addition, the company also has the technology to use reclaimable silicon that acts as a further cushion in terms of raw material supplies. (Investor Presentation, September 2008).

Of course, it is possible that poly prices will decline once significant additional capacity comes online in the next two years, in which case companies that have not locked in supplies could be greatly benefited. However, an investor looking at such companies would have very little visibility on poly procurement costs going forward.

Another factor influencing supply visibility for a crystalline silicon company is the track record of its poly suppliers. For instance, some of Evergreen's suppliers, such as DC Chemical, do not have a long track record in manufacturing polysilicon. China Sunergy has faced delays and failures from its suppliers several times in 2007. Yingli Green has had to cancel (and replace through new contracts and in the spot market) supplies of poly amounting to ~19 percent of contracts signed in 2007 because suppliers failed to deliver on their contracts. The weaker the track records of suppliers, the lower the supply visibility.

SUPPLIER CONCENTRATION

A solar company that is heavily dependent on a few major suppliers for critical raw materials or equipment is more exposed to supplier risk than are companies with a large number of suppliers. For example, in its 20-F filing for FY07, JA Solar notes that since some of its manufacturing equipment is uniquely tailored to its specific needs, it may not have a choice of seeking alternative suppliers within a short lead time. Companies like Evergreen Solar and China Sunergy face a similar issue with custom-made capital equipment.

For China Sunergy, the top 10 suppliers accounted for ~55 percent of silicon raw material supply in FY07. Further, most of these were under contracts of less than a year in duration, creating issues in terms of visibility.

Yingli Green also has a high degree of supplier concentration. In FY07, its top five suppliers provided more than 70 percent of its polysilicon supplies.

Evergreen Solar currently uses a special form of string for its unique string ribbon manufacturing process, and it sources this from a single supplier. However, the company is taking steps to produce this string at its Devens facility in order to have better control over a crucial raw material in its manufacturing process.

Similarly, First Solar is dependent on a limited number of suppliers for most of its raw materials, including cadmium telluride, a critical raw material in its thin film process. Evergreen Solar also depends on a limited number of suppliers for key inputs.

In contrast, Suntech Power sources polysilicon/silicon wafers from more than 40 different suppliers, giving it a much more diversified supplier base. Even if one or two of its suppliers were to face disruptions in their operations, Suntech would still be able to carry on largely unhindered.

An investor would feel a lot more comfortable investing in solar companies that have a well-diversified supplier base, which reduces supply-side risk and also gives the company greater negotiating power with suppliers. Having a diversified supplier base is particularly important in the solar sector because several of the poly suppliers are new entrants with no significant track record, which makes it risky to rely on just one or two suppliers.

DEMAND VISIBILITY

An investor would feel more comfortable investing in a solar stock if the demand visibility is high. During the period of polysilicon shortage, some solar companies negotiated multiyear, "take or pay" contracts with customers (mainly project developers/system installers/distributors) of solar modules. Their strategy was virtually the mirror image of the polysilicon companies. The result was a substantial hardening of the supply chain. Arguably, the high barriers to entry erected by long-term take-or-pay contracts resulted in an identifiable allocation of raw materials (polysilicon) and modules that drove up valuations during the period. Recall, however, that demand did not turn out to be "endless," as rising inventory and dramatic price declines proved exiting 2008.

Evergreen Solar has signed take-or-pay contracts with six customers, adding up to a cumulative value of almost $1 billion through 2011 (out of which $170 million was fulfilled by the end of December 2007). First Solar has contracts with 12 European customers for 3.2 GW of solar modules representing revenues of ~$5.9 billion from 2008 to 2012. While these contracts likely have price declines built in, their take-or-pay nature may not withstand the market realities, which could lead these companies to renegotiate or otherwise cancel these contracts should they become uneconomical for the buyers.

Good demand visibility enables the investor to make a more informed call on the performance of the company and the stock.

One of the factors influencing demand visibility is the percentage of revenues that comes from retail versus commercial customers. If a large share of a company's revenue comes from retail customers, we would expect delivery lead times to be shorter and demand visibility to be lower. This appears to be the case entering 2009.

SunPower stated at its analyst day held on November 11, 2008, that the credit crisis has led it to "shift gears" into the residential market for cash deals and to value added resellers for the same reason that large project financing has essentially put the project market growth rate on hold. This is truly strength for SunPower, as it may shift into various channel strategies to best accommodate the changing market. Other makers that do not have a multichannel strategy will need to consider what to do if their long-term contract customers, dependent on reasonable or even existent bank loans to build their projects, come back to renegotiate lower prices, delayed deliveries, or both.

CUSTOMER CONCENTRATION

A company heavily dependent on a few major customers is more exposed to the risk of losing one or more of its top customers and would also have less bargaining power with its customers. Therefore, an investor would want to know the extent of this risk.

For example, in FY07 more than 40 percent of revenue for JA Solar came from its top three customers, and more than 75 percent of its revenues came from its top 10 customers.

Similarly, in FY07 each of the top six customers of First Solar accounted for 10 to 23 percent of its revenues, while none of the other customers accounted for more than 10 percent of revenues. Moreover, all of the top six customers of First Solar were based in Germany, exposing the company to a high degree of geographic risk as well.

For Energy Conversion Devices, ~40 percent of FY07 revenues came from its top five customers. For Trina Solar, ~49 percent of FY06 revenues came from just the top five customers.

An investor would ideally want to invest in a solar company with a broad, diversified customer base, so that the company is not excessively dependent on a few large customers. Also, a company with a geographically diversified customer portfolio would present less country risk to an investor.

TRACK RECORD IN ACHIEVING SCALE

It is important for a solar company to be able to move technology from the laboratory to the factory as quickly as possible. The process of manufacturing PV products is intricate, and the success or failure of a ramp could have a substantial impact on a company's finances and stock price. As Yingli Green notes in an SEC filing, "Deviations in the manufacturing process [for PV modules and systems] can cause a substantial decrease in output and, in some cases, disrupt production significantly or result in no output." This is true not just for Yingli Green but for all companies in the solar industry. If a company has a reliable approach to replicate its production lines in order to ramp up, it would be able to achieve scale with more ease and reliability. For example, First Solar has used a "standard building block" approach wherein its production lines in Ohio were used as a template to build the lines in its German factory.

Energy Conversion Devices presents an interesting case where ramp times have decreased dramatically over the last few quarters in 2008. With the ramp-up at its Greenville facility proceeding ahead of schedule, the company expected to increase capacity by ~140 percent year over year to around 25 MW by the end of 4Q08. As execution risk may have several or many underlying causes, a company's ability to scale commercially is a significant risk and must not be underestimated.

As mentioned earlier, thin film technologies have a scalability advantage in the sense that the deposition techniques they employ can be more easily moved from the laboratory to the factory.

Suntech Power plans to manufacture higher-efficiency solar cells using the Pluto technology and has in place a pilot line to achieve conversion efficiencies of 18 to 19 percent. However, it is important for Suntech to successfully create a large scale production line to deploy this technology. Given its relatively limited operating history (Suntech started operations only in 2002), an investor would want to know how feasible it is for the company to achieve the required scale. Similarly, JA Solar also has a limited operating history (it completed its first solar cell manufacturing line in

December 2006), which means an investor cannot use a track record as a sufficiently reliable predictor of the future.

Despite the newness of many solar companies, the investor must judge execution risk by allowing for increases in his discount rate for either or both new companies and new technologies.

MANAGEMENT QUALITY AND CORPORATE GOVERNANCE

Management quality is a parameter, which, though subjective and qualitative, is a crucial element of an investor's decision parameters. Given that the solar technology industry is growing so quickly, it is imperative for an investor to determine whether a company's management has the credentials and ability to manage this growth effectively.

Corporate governance is another area where investors seek reassurance. An investor would like to feel confident that the numbers reported by the company accurately reflect its performance. He or she would also want to know if there are any potential pitfalls related to conflicts of interest.

We analyze these critical issues along three lines, which we will go into below.

1. Management's Experience in the Company and Industry

This is a simple yet reliable indicator of a company's ability to deliver on its plans. Given the emerging nature of the industry, management personnel with experience in technology-and-manufacturing-intensive businesses such as the microprocessor industry could also be considered as having excellent credentials for the solar technology industry. A few examples will describe how the management of a solar company could be evaluated.

Top management at JA Solar includes several officers who are very new to the company. The president and chief operating officer, chief technology officer, and chief financial officer have been with the company for less than a year each.

Management at SunPower has been stable over the last five years, with key management personnel having spent quite a few years with the company. For example, the CEO, Thomas Werner, has held the position for over five years, and the president and CTO, Dr. Richard Swanson, has been with SunPower for over 20 years. Other key management personnel also come with strong experience and credentials in the solar industry.

Evergreen Solar also has an experienced management team. President and CEO, Richard Feldt, for instance, has been with Evergreen for over four years and has significant experience in managing technology-intensive businesses.

Energy Conversion Devices has leading photovoltaics experts such as Dr. Subhendu Guha and Michael Fetcenko in its management ranks. On the other hand, several of the other management personnel—including the president and CEO, Mark Morelli—are relative newcomers to the company and industry. However, we can trace this to attempts at bringing Energy Conversion Devices to profitability and to improving its operations (in fact, the company did turn profitable in 3QFY08).

The management team members at First Solar are generally composed of personnel experienced in high-technology industries. However, an analysis of their tenure with the company presents a mixed picture: while the CEO, Michael Ahearn, has held his position for nearly eight years, many of the others have been in their positions for just one to two years.

Such an analysis of the company management's credentials and track record would lend strong support to an investor's decision on whether to invest in the company.

2. Internal Controls

The presence of effective internal control mechanisms for financial reporting reassures an investor about the transparency and correctness of a company's reported numbers. An investor can gauge this from the report of the company's public accountants. Two contrasting examples illustrate this. JA Solar's public accounting firm has identified a major gap in its internal controls: the lack of adequately qualified personnel to handle accounting according to the United States GAAP. If this situation is not redressed quickly, the company's credibility could be undermined in the eyes of investors. On the other hand, Energy Conversion Devices has been certified by a public accounting firm as having effective internal controls.

3. Potential Conflicts of Interest and Other Corporate Governance Issues

An investor would want to know up front about any conflicts of interest that senior management members might have. For example, JA Solar's chairman is also the chairman of one of its major suppliers, Jinglong BVI. This represents a potentially significant conflict of interests.

Further, some companies have antitakeover measures incorporated in their charter, which could have a bearing on their stock price. Evergreen Solar, for instance, has instituted measures in its charter and by-laws to dissuade takeover attempts; it also has a staggered system of electing its board of directors, which makes it difficult for a hostile bidder to replace the board in a single stroke. An investor in a solar company would want to look at these provisions and study their potential impact in the future.

In the case of First Solar, a significant percentage of shares (>45 percent) is held by a single shareholder and affiliates, who therefore have a strong voice in significant company-related decisions. A concentrated shareholding distribution may make a company's stock less attractive to investors due to the possibility of a majority of shareholders taking or supporting actions that might undermine the rest of the shareholders.

EVERGREEN SOLAR—THE LONG-TERM VIEW

This discussion of Evergreen Solar's long-term prospects is excerpted from research produced by Alternative Energy Investing (AEI).

The Cost Structure Question

Evergreen Solar (ESLR) can rightfully point to its highly efficient use of polysilicon as a key competitive advantage. At roughly five grams/watt, ESLR uses 50 percent less polysilicon than most other manufacturers. However, its capital equipment costs are higher than average, which erodes the benefits of the company's polysilicon efficiency. The critical question for ESLR is whether and when its cap equipment costs will fall into line with those of other crystalline silicon cell producers to reveal the financial benefits of string ribbon technology. We believe they will, but the process may take at least two years.

The Furnace Factor

ESLR's cell and module capital equipment set is mostly the same as that of other crystalline silicon cell and module producers. The crucial differences are in the silicon growth and wafering steps. ESLR's ribbon silicon growth furnace and laser scribe equipment cost roughly 30 percent more than the standard crystalline silicon equivalent (an ingot furnace and silicon saw). Because of the nature of ESLR's ribbon growth technology, its solar cells (at three inches) are only half as wide as the industry standard (six inches). This nonstandard cell size adds a small increment to the company's cell processing costs.

Quantifying the Difference

We believe the incremental cost of the furnace and scribe equipment have only a minor effect on ESLR's cost structure and is more than recouped by the improvements in polysilicon use per watt of production. If, for example, ESLR's furnaces and scribes represent 50 percent of its equipment cost per MW of capacity, then ESLR's all-in capital cost is 15 percent higher than the industry average. The implications of the higher capital cost are modest—the 15 percent capacity cost premium translates into a 3 percent higher depreciation charge per year (based on a five-year depreciation schedule) applied to ESLR's cost of solar module production. Meanwhile, the company's superior polysilicon utilization translates into about 13 percent cost advantage per watt at the module level, given "normalized" polysilicon prices (see Tables 7.3 and 7.4). The net effect is a potentially sustainable cost advantage of roughly 10 percent, a critical margin in a commodity industry.

We now highlight the key assumptions used to derive the numbers in Tables 7.3 and 7.4.

- "Base case" price of $125/kg represents AEI's estimate of the solar industry's current average price paid for polysilicon.
- "Normalized price" of $50/kg represents AEI's opinion of the stable long-term price of polysilicon given the cost structure and return requirements of the polysilicon industry.

TABLE 7.3

Evergreen Solar's Cost Structure at Current Polysilicon Usage Rates

Polysilicon price	Normalized Price $50/kg		Base Case $125/kg		Shortage Price $200/kg	
	Industry	**ESLR**	**Industry**	**ESLR**	**Industry**	**ESLR**
Grams of polysilicon/watt	10.0	5.0	10.0	5.0	10.0	5.0
Polysilicon cost/watt	$0.50	$0.25	$1.25	$0.63	$2.00	$1.00
Module costs ex-polysilicon	$1.38	$1.38	$1.38	$1.38	$1.38	$1.38
Total module costs	*$1.88*	*$1.63*	*$2.63*	*$2.01*	*$3.38*	*$2.38*
ESLR cost vs. industry	−13.3%		−23.6%		−29.6%	

Source: Alternative Energy Investing

TABLE 7.4

Evergreen Solar's Cost Structure at Projected Polysilicon Usage Rates

Polysilicon price	Normalized Price $50/kg		Base Case $125/kg		Shortage Price $200/kg	
	Industry	**ESLR**	**Industry**	**ESLR**	**Industry**	**ESLR**
Grams of polysilicon/watt	7.0	2.5	7.0	2.5	7.0	2.5
Polysilicon cost/watt	$0.35	$0.13	$0.88	$0.31	$1.40	$0.50
Module costs ex-polysilicon	$1.75	$1.75	$1.75	$1.75	$1.75	$1.75
Total module costs	*$2.10*	*$1.88*	*$2.63*	*$2.06*	*$3.15*	*$2.25*
ESLR cost vs. industry	−10.5%		−21.7%		−28.6%	

Source: Alternative Energy Investing

- "Shortage price" of $200/kg represents the polysilicon contract price paid by Evergreen Solar during the recent polysilicon supply shortage. AEI notes that polysilicon contract pricing has already eased and that Evergreen's most recent multiyear polysilicon contracts appear to be priced in the $50 to $70 range.

- "Module costs ex-polysilicon" are based on AEI's estimated direct cost of solar module production of roughly $2.63 per watt.
- "Grams of polysilicon/watt" for Evergreen Solar and the industry are based on company guidance and AEI estimates.

Evergreen Solar has also articulated its growth strategies for the future, which are presented below.[1] These could be useful to the long-term investor in terms of understanding where growth may come from.

Achieve Costs Low Enough to Attain Grid Parity

This will be done both through reductions in manufacturing costs and through reductions in other parts of the value chain. For example, Evergreen Solar has relationships with utilities, installers, and other entities that help reduce installed system costs and levelized costs of electricity.

Manufacturing-Related Innovations

As we have seen, Evergreen Solar's success has been heavily dependent on its unique manufacturing process. As the company grows, it intends to maintain its leadership in this area. Evergreen Solar expects to be able to achieve conversion efficiencies of ~18 percent and factory yields of ~90 percent by 2012.

Increasing Fully Owned Capacity

Evergreen Solar is a partner in EverQ, a joint venture with Renewable Energy Corp. and Q-Cells, which involves significant capacity additions. At the same time, the company also intends to increase its fully owned manufacturing facilities so that it will be able to derive maximum value out of its proprietary string ribbon technology.

Potential Entry Points for New Investors

On July 8, 2008, Evergreen Solar announced the closing of its $374 million offering of senior convertible notes, out of which its ultimate proceeds were $322 million. These proceeds were meant for

(1) the construction and other activities related to its solar panel facility in Devens, Massachusetts; (2) manufacturing facilities to produce special heat-resistant string (for its proprietary string ribbon process), and (3) other corporate purposes.

A few months earlier, on February 15, 2008, the company had announced the closing of a public offering of common stock resulting in net proceeds of ~$166.9 million to Evergreen Solar. This was to be used for the completion of phases one and two, noted above, of its Devens manufacturing facility and general corporate purposes. Such offerings could serve as good entry points for new investors.

COMPANY EXAMPLES FOR A BETTER UNDERSTANDING

First Solar: Lowest Manufacturing Costs

Phoenix, Arizona, based First Solar (FSLR) makes thin film solar cells and modules with cadmium telluride semiconductor material on a glass substrate. What makes First Solar stand out to investors is that it has the lowest manufacturing costs per watt ($1.18/W as of 2Q08) in the industry. The low manufacturing cost structure means FSLR has solid gross margins of 50-plus percent even with low ASPs ($2.57/W in 2Q08). It also makes FSLR modules highly attractive for large scale, commercial PV installations.

Technology Advantage

As noted, FSLR's modules are based on cadmium telluride, which enables the modules to achieve good conversion efficiencies while using just a fraction (around 1 percent) of the semiconductor material required by c-Si cells. Indeed, as of 2Q08, First Solar has been able to achieve conversion efficiencies of 10.7 percent using its cadmium telluride cells.

Power output from solar cells generally decreases as cell temperatures increase. That is to say, as the ambient temperature increases, power output decreases. With cadmium telluride cells, however, performance does not deteriorate as much as in the case of crystalline silicon cells as temperatures increase. Further, cadmium telluride cells are also better at converting low light into electricity.

Manufacturing Geared to High Volumes

Apart from these advantages offered by use of cadmium telluride material, First Solar enjoys several benefits from the manufacturing perspective. The production process is geared to high-volume, continuous manufacturing.

Further, First Solar uses a "Copy Smart" methodology that allows it to rapidly replicate manufacturing facilities. To give an example, Copy Smart was successfully adopted to build First Solar's recently completed lines in Malaysia. Indeed, the company expects to leverage this to take production to >1 GW by the end of 2009.

Cost Advantage and Focus on Large Scale PV

As mentioned earlier, the clear advantage on per-watt costs offered by First Solar means that it is a particularly attractive choice for commercial PV installations. At manufacturing costs of $1.18/W as of 2Q08, FSLR's costs are far lower than those of crystalline silicon players. In fact, even under a normalized silicon price scenario of $50/kg, and with reduced silicon consumption of 7g/W, crystalline silicon manufacturers can achieve costs of approximately $1.35/W. If by that time FSLR can manage to bring down its costs to $0.75/W, it would mean a cost advantage at the module level of ~44.5 percent per watt. Even allowing for some reduction of this advantage at the system level, FSLR can still be expected to maintain a significant cost advantage over its silicon-based rivals, maintaining its attractiveness in large scale deployments of PV.

Further, as cost reductions continue, First Solar seems well poised to reach grid parity, at which point it will be able to compete with other sources of electricity without subsidies. In fact, First Solar has been targeting cost reductions so it will be able to price modules at ~$1.00 to $1.25/W some time between 2010 and 2012. With a corresponding decrease in balance-of-system costs, First Solar expects the total systems cost to be ~$2.00 to $2.50/W, driving down the levelized cost of electricity to around 8 to 10 cents per kilowatt hour sometime between 2010 and 2012. This would enable FSLR's PV modules to become cost competitive with other sources of electricity.

We also note that in keeping with its focus on commercial scale PV installations, First Solar does not sell its modules through resellers.

Instead, it has tie-ups with a few system integrators capable of installing large scale PV systems.

Q-Cells: Technology Diversification

Founded in 1999, Q-Cells is one of the largest solar cell manufacturers globally. The core business areas of the company include development, production, and sale of mono- and multicrystalline silicon solar cells. Apart from this, Q-Cells also develops and produces thin film modules using different technologies.

Q-Cells' position in the solar PV supply chain is illustrated in Figure 7.1.

What makes Q-Cells stand out to investors is that it has not only a strong market growth with a proven profitability track record, but also a portfolio of innovative, advanced products built on diverse technologies.

Diverse Technology Advantage

Not only is Q-Cells a major player in silicon-based PV, but it also has a unique portfolio of new technologies under its umbrella. In the wafer-based technology, Q-Cells has polycrystalline silicon technology, monocrystalline silicon technology, string ribbon technology, and low concentration PV technology. In the thin film business, the company uses two types of substrates: fixed substrates (glass) and flexible substrates. The technologies in the fixed substrate segment are micromorph/tandem silicon, cadmium telluride, CIGS, and crystalline silicon on glass (CSG). The company has amorphous-silicon-on-plastic-foil technology (flexcell) under the flexible substrates segment.

Q-Cells has several strategic partners where the company has a high percentage of ownership. These partnerships enhance the

Source: Q-Cells AG

Figure 7.1 Position of Q-cells in the supply chain of photovoltaics

technological diversity that makes the company an attractive choice in the long term.

Examples of strategic stakes acquired by Q-Cells in the wafer-based technology include: Renewable Energy Corporation (17.18 percent), EverQ (33.33 percent), and Solaria (31.4 percent). Examples of strategic stakes acquired by Q-Cells in the thin film business under the fixed substrates segment include: Sontor (100 percent), Calyxo (93 percent), Solibro (67.5 percent), and CGS Solar (21.71 percent). Q-Cells also has a 57.1 percent partnership with VHF-Technologies in the flexible substrates segment.

The company maintains a strong focus on research and development. Currently, it is involved in the development of next generation high-efficiency cell concepts.

Good Growth Prospects

Q-Cells has good revenue visibility and more than 90 percent of 2009 production volumes already sold. As part of its growth strategy the company will undertake a plant and capacity expansion, augment wafer supply, and produce high value-added products/brands. It has a target of more than two GW_p of production volume by 2010. Also, Q-Cells is aiming at a customer base expansion along with further business and geographical diversification.

Q-Cells has already secured a silicon and wafer supply for strong long-term expansion. The company aims to have more than 30 percent of its total silicon supply from upgraded metallurgical silicon by 2009, and around 50 percent by 2010.

Significant Push into Upgraded Metallurgical Silicon

The company has planned a major push into UMG-Si as part of its effort to reduce silicon costs. In July 2008, Q-Cells signed a contract with BSI/Timminco for upgraded metallurgical silicon/solar grade silicon (directly purified metallurgical silicon). This calls for supply volumes of 410 tons in 2008, 3,000 tons in 2009, and 6,000 tons per year from 2010 till 2013. Q-Cells will develop and make use of in-house know-how to process the upgraded metallurgical silicon. It will set up a center of expertise with 25 MW_p capacity in (Germany), and integrate ingot and wafer production into its manufacturing complex in Malaysia. The Malaysian complex is expected to be operational and increase the total cell production capacity to 1030 MW_p by mid-2009.

Given the higher level of impurities in upgraded metallurgical silicon, backward integration into ingots and wafers would enable Q-Cells to maintain the requisite control to ensure the quality of its cells.

Other related areas on which the company is working include technological advancements such as reduction of cell thickness, increase of cell efficiency, global expansion (production complex in Malaysia), and achieving economies of scale.

Risks for the Long-Term Investor

Q-Cells is technologically very diverse and thus does not face a significant degree of technology risk. Operationally, however, it faces risks from its foray into upgraded metallurgical silicon, since this has not been extensively tested on a large scale before.

SunPower: A Different Business Model

SunPower has two business segments, Systems and Components. The Components business involves selling PV modules. The Systems segment, which is an installation business, contributes up to 60 percent of the total business revenue. It mainly comprises the PowerLight business acquired from SP Systems in January 2007. Globally, more than 400 SunPower solar power systems are commissioned or in construction that are rated in aggregate at more than 300 megawatts of peak capacity. SunPower is also entering the financing of the solar PV Systems segment.

To give an example of the kind of activities undertaken by the Systems business, in December 2007 the company entered into an engineering, procurement, and construction agreement (the "EPC Agreement") with the Spanish company Solargen Proyectos. According to this agreement, SunPower would design and construct a PV plant of ~8.3 MW_p in the Extremadura region of Spain.

Recently, SunPower also signed a 250 MW agreement with Pacific Gas & Electric (PG&E) through third-party power purchase agreement counterparty High Plains Ranch II, LLC, SunPower's own financing entity. This enabled SunPower to enter into an arrangement where it will sell only the solar PV electric output and not transfer the ownership of the system itself.

These are some of the attempts by SunPower to convert its business model into an innovative integrated business model, to leverage and enhance its depth of customer offerings. In other words, SunPower will be able to integrate production, installation, and financing of the solar PV systems. This could prove a very attractive long-term advantage for the company, but on the other hand could add the risk of a heavy balance sheet for the company, along with lower Systems margins.

LDK: Vertical Integration

LDK Solar started out as a wafer manufacturer, with 420 MW of wafer capacity by the end of 2007. However, the company is currently in the process of integrating backward into polysilicon manufacturing as well. In fact, LDK had initial plans to set up an installed capacity of a massive 16,000 MT of poly by the end of 2009, with actual production of 5,000 to 7,000 MT of poly planned in 2009. Because of the economic slowdown, LDK is delaying its capacity expansion plan. This poly capacity is mainly intended as feedstock for its own wafer operations.

LDK expects poly produced from its plant to cost less than \$32/kg. Further, the poly produced is expected to be of very high (11 N) purity. LDK's backward integration plan is in line with its wafer capacity expansion plan: the company intends to increase wafer capacity to 2.0 GW by the end of 2009, which was lowered from the 2.3 GW previously forecast.

On a different note, LDK also has plans for upgraded metallurgical silicon-based wafers. Indeed, LDK has started shipping its "Nova" wafers using upgraded metallurgical silicon, which result in efficiencies of 13 to 15 percent.

LDK's growth strategy is built on the following areas (LDK Solar 2008 Analyst Day, July 16, 2008, San Francisco, slide 14):

- Ensuring polysilicon supply from third-party vendors and recycling
- R&D to improve both technology and manufacturing operations

- Partnerships with other companies and possibly inorganic growth
- Backward integration into poly and use of upgraded metallurgical silicon
- Geographic diversification
- Wafering: capacity expansion and quality enhancement

From the perspective of a long-term investor, backward integration has potentially substantial benefits, but it also comes with risks. As noted earlier, each process in the solar PV production is extremely specialized, and success is by no means guaranteed. In summary, a long-term investor in LDK could stand to gain if the company manages to execute its backward integration strategy without major problems in operations.

Trina Solar: Vertical Integration

Trina Solar also follows a vertical integration strategy, with a presence across silicon ingots, wafers, cells, and modules. According to the company, vertical integration creates the following advantages (Trina Solar, investor presentation, September 2008, slide 14):

- Allows Trina Solar to capture more value across the solar supply chain. In fact, Trina's gross margin increased from 22.4 percent in 1H07 to 24.2 percent in 1H08 partly as a result of its vertical integration strategy. This is particularly noteworthy given that silicon raw material supplies in this period remained tight.
- Enables better feedback and therefore quality control.
- Allows for better yields and conversion efficiency.
- Reduces the amount of time required for product development.

As it pursues a vertical integration strategy, Trina Solar intends to increase focus on the cell and module portions of the supply chain in order to capture more value. Indeed, by year-end 2009, Trina Solar plans to have more capacity in cells and modules, as compared to ingots and wafers, so that modules would become the main source of revenue.

The company highlights the following as its core strengths (Trina Solar, investor presentation, September 2008):

- It has invested in a flexible manufacturing model. This in turn gives the following benefits:
  - Allows better use of capital
  - Enables the company to penetrate the market faster
  - Allows the company to use the supply-demand situation across different parts of the value chain to its advantage
- It has a track record of executing its capacity expansion plans.
- It is well-positioned in terms of its raw material requirements, with 80 percent of raw materials secured for a planned 450 MW output in 2009 as part of long-term contracts. Long-term contracts include agreements with GCL Silicon, DC Chemical, Nitol, Wacker, and Silfab.
- It has an established brand, which is very important in the solar market space.
- It has a global base of customers and also a number of well-established and reputed customers that add to its credibility.
- The company has also been investing in improving its technological capabilities.

Trina Solar has also been trying to diversify its geographic concentration of revenues, as evidenced by the progression of revenue mix from what it was in 2007 to what is expected in 2009:

- 2007: 40 percent Spain, 31 percent Germany, 18 percent Italy, 7 percent other countries in Europe, 2 percent China, rest other countries
- 2009 (expected): 28 percent Germany, 25 percent Italy, 15 percent Spain, 8 percent United States, 8 percent France, 7 percent Belgium, 4 percent Korea, rest other countries

As these numbers indicate, Trina Solar's revenues were geographically very concentrated in a few countries in 2007, but are forecast to be much more diversified geographically in 2009.

A long-term investor may see an attractive stock based on Trina Solar's vertical integration strategy and its execution capabilities, and also some of the other areas described above (raw material supplies, flexible manufacturing model, geographic diversification of revenue, etc.).

INVESTING ON THE BASIS OF REVENUE AND EARNINGS GROWTH

The idea behind investing in revenue and earnings growth is based on the belief that if a company has witnessed strong revenue and earnings growth over the last few years or last few quarters, it might do so over the coming few quarters or years as well. Historical revenue and earnings growth point to the ability of a company to scale up operations and generate value for shareholders. Future revenue and earnings growth can be estimated approximately from projections of the company's capacity additions, estimated average selling prices, and costs per watt. A higher expected revenue and earnings growth would justify higher valuations because of the potential for higher cash flows, which would be discounted into the stock.

An investor could estimate a solar PV company's potential for revenue and earnings growth from the following:

- Its plans and ability to scale up and ramp capacity. For example, if a quality company like Suntech Power announces a capacity expansion schedule, investors can feel reasonably assured of the potential for revenue and earnings growth.
- Its ability to maintain or improve ASPs. If a company's solar modules have differentiating features, it may also have the ability to command higher ASPs by virtue of these unique features (for example, SunPower has better aesthetics; Energy Conversion Devices Uni-Solar laminates are lightweight and easy to install for building integrated photovoltaics applications). ASPs are very important because they enable a solar PV company to generate superior revenues and earnings.
- Its ability to keep strict controls over costs. If a company has excellent control over its costs, it will have better margins and therefore superior earnings; for example, First Solar, which has the lowest per-watt manufacturing costs in the industry and (as of Q2 2008) gross margins of 50-plus percent.

STRATEGIC RISKS

Risks Related to Business Models

Solar PV companies could be exposed to risks as existing business models change and are replaced with new ones. We are already seeing this with PV companies beginning to get into the financing space as well. Other innovative business models could emerge that might threaten existing ones.

MACROLEVEL RISKS

Subsidy Related Risks

Before we analyze this risk, we need to make a set of assumptions that the world economy remains robust and doesn't go into a recession, that fossil fuel prices (coal, gas, and oil) will not decline substantially as a result of declining global demand or excessive generation of renewable energy by different technologies, and that there is no serious risk from higher global interest rates.

Solar is still an expensive form of energy, and hence its demand is directly related to subsidies, incentives, and various policies for solar PV. Also, solar demand is dependent fundamentally on the world markets, so any policy changes in the subsidies would have an immediate impact on the solar industry. For instance, the current focus of subsidy programs in governments across the globe is on BIPV rather than ground-mounted solar. As the support for the solar PV technologies increases, the energy costs will decline.

Credit Risks

Many solar companies make advance payments to their polysilicon and silicon wafer suppliers, and the credit term sales offered to some of the customers expose these companies to the credit risks of such suppliers and customers. This may increase the solar company's costs and expenses, which could in turn have a material adverse effect on the liquidity.

Solar companies make advance payments to suppliers prior to the scheduled delivery dates for silicon wafer supplies. In many such cases, the advance payments are made without receiving collateral.

The solar company may also offer short-term and/or medium-term credit sales based on the relationship with certain customers and market conditions, also in the absence of receiving collateral. This exposes the company to credit risks in the event of insolvency or bankruptcy, since the claim for such payments or sales credits would rank as unsecured claims. This could have a huge negative effect on the company's financial condition, results of operations, and liquidity.

Here are a few examples of credit risks faced by solar PV companies:

- As of the end of December 2007, Renesola had made payments of ~$53.7 million to its suppliers in the form of advances.
- SunPower has entered into agreements with Hemlock Semiconductor where it will make prepayments to the tune of $113 million by 2010.
- As of the end of 2007, LDK Solar had made prepayments of ~$157.2 million to suppliers.

Most other solar PV companies also have similar prepayment agreements with their silicon suppliers that could expose them to significant credit risks.

OPERATIONAL RISKS

Risks in Capacity Expansion and Cost Reduction

Capacity expansion can also be risky for a solar PV company. A lot of care needs to be taken in ramping and maintaining operations so that even a small error is avoided, since even that can drastically reduce solar PV efficiency.

Also, as most companies try to reduce the costs by using different technologies, such as upgraded metallurgical silicon, this increases the risk. Upgraded metallurgical silicon, for instance, is a new technology and has not been tested extensively till now, and heavy exposure and reliance on it might increase risk. Q-Cells and Canadian Solar, for example, are becoming increasingly reliant on upgraded metallurgical silicon for the continued success of their operations. While they would certainly have tested upgraded metallurgical silicon considerably before deciding to adopt it, it is not certain that the same yields and efficiencies would be obtained during commercial, large scale cell production using upgraded metallurgical silicon.

To tackle the expected dearth in polysilicon supply, many companies are also getting into backward integration of their business; this also increases the risk, as each stage in silicon manufacturing is specialized and the company has no experience in the new process.

Solarfun, for example, has a vertical integration strategy that carries with it significant operational risks. It has acquired a company called Yangguang Solar in an attempt to bolster ingot supplies. However, Yangguang Solar is a new entrant with little experience in managing massive ramp-ups within short durations. Given that the PV industry is complex in terms of manufacturing, execution of Solarfun's strategy comes with potential risks.

Risks Related to Funding Expansions

Several fast-growing solar PV companies plan to increase capacity on a massive scale. This means a high requirement for capital. Assuming a $1/W capital cost, a 500 MW expansion would require access to $500 million. A company may need external funds to finance such expansions due to their weak balance sheets. But this involves a risk, as many banks are suffering due to the losses they incurred on subprime real estate portfolios. This could hamper their ability to lend money in the form of loans or credits for solar energy projects to utilities. Even if the banks continue to lend money, it might be at an elevated interest rate. This can have a negative effect on the solar PV capacity expansions and could ultimately mean a reduction in demand for PV installations.

PRODUCT-RELATED RISKS

Health Risks

First Solar uses cadmium telluride extensively in its solar PV cell. The cadmium telluride compound semiconductor material forms the active PV cells, which convert sunlight into electricity.

Cadmium telluride is highly toxic and very harmful to human beings who come in direct contact with it.

The PV modules have been created in a way that excludes the possibility of direct contact with cadmium telluride, unless the PV system is exposed to fire: but fires in residential and commercial properties are not uncommon. If such fires involve the roof, PV arrays mounted there will be exposed to the flames, and cadmium will be released, which could lead to serious health risks. And there may be other unforeseen circumstances that might result in leakage of cadmium telluride. Even isolated incidents could potentially trigger adverse reactions toward the product, and therefore the stock price.

Environmental Risks

The disposal and long-term safety of the poisonous cadmium telluride is an issue in the large scale commercialization of cadmium telluride solar PV panels. It can prove to be a big environmental hazard if not handled properly. This is a huge risk for First Solar, which uses cadmium telluride extensively.

Performance/Warranty Risks

Another major risk is the lifetime of the modules. For instance, First Solar claims that its modules have a 25-year warranty. Though the modules are subjected to extensive testing that replicates the environmental conditions, this could still be a risk. If the life span of the modules is less than what has been anticipated or the performance is less than what has been expected, the installations would be less profitable for the utilities and other commercial setups.

COST REDUCTION IN THE INDUSTRY

In an industry where products would tend to become increasingly commoditized, cost reduction is an area to which all solar companies pay a great deal of attention. In general terms, cost reduction could be achieved in several ways:

1. Shifting to alternative, lower cost raw materials
2. Reducing the amount of raw material required per watt

3. Reducing equipment costs and therefore capital costs

4. Better sourcing agreements with raw material suppliers; entering into contracts with multiple suppliers to obtain better bargaining power

5. Shifting manufacturing to low cost locations, resulting in lower capital costs as well as operating costs

6. Achieving manufacturing efficiencies through process improvements

7. Technology improvements resulting in better conversion efficiencies and therefore lower per-watt costs

We now look at specific company examples to illustrate how cost reductions might be achieved.

Energy Conversion Devices

Energy Conversion Devices has a road map to achieve grid parity by 2012 (Energy Conversion Devices, presentation at PiperJaffray Clean Technology and Renewables Conference, February 20, 2008). The road map requires cutting levelized costs of electricity from 21 cents/KWh to around 8 to 10 cents/KWh. This in turn would require a reduction of $1.47/W in panel costs to reach $1.10/W by 2012. Energy Conversion Devices plans to achieve this by simultaneous efforts on three fronts. If the company's efforts proceed as planned, these three would eventually contribute one-third each to the cost reduction initiative.

1. *Material sourcing.* As Energy Conversion Devices ramps production, it would require raw materials on a much larger scale than it does currently. This would give the company better leverage in negotiations with suppliers. The company also plans to rope in new suppliers in an effort to reduce costs of raw materials such as stainless steel and grid wire.

2. *Ramp-up and manufacturing related improvements.* Energy Conversion Devices is ramping up capacity faster than anticipated, Greenville 1 being an example. To reduce costs, the company would expect to maintain the ahead-of-schedule ramps. Also, with improved cell quality being achieved on the manufacturing lines, costs can be brought down further.

3. *Conversion efficiency and other improvements.* As mentioned earlier, improved conversion efficiencies can also help reduce per-watt costs. Energy Conversion Devices has been increasing production and shipments of its 144-watt product, which has a conversion efficiency of 8.5 percent, as against the ~8 percent conversion efficiency of the rest of its product line. Also, its efforts to include low cost locations (like Tijuana, Mexico, and Tianjin, China) as part of its manufacturing strategy led to further potential cost reductions.

Suntech Power

Suntech Power uses conversion efficiency as a key lever to reduce costs. Using Pluto technology, it has achieved efficiencies of 17 to 18 percent, with the cost of production per wafer essentially remaining intact. Further, Suntech Power also expects to be able to use thinner wafers in the future on the Pluto platform, which would mean a decrease in grams per watt and therefore the cost per watt of silicon. Besides these, Suntech Power also has the advantage of economies of scale, considering its large scale capacity expansions, which would further help lower per-watt costs.

Evergreen Solar

Evergreen Solar is already the benchmark in the industry as far as polysilicon usage is concerned, with less than 5 grams of silicon required per watt as compared to the industry average of ~10 grams per watt. As mentioned earlier, this advantage with regard to silicon consumption arises from Evergreen Solar's proprietary string ribbon technology.

Evergreen expects to maintain this silicon cost advantage several years into the future. While the industry would reach 7g/W of silicon by 2014, Evergreen Solar expects to cut consumption to just 1.5g/W (Evergreen Solar Capital Markets Day presentation, June 19, 2008). The company has also developed an "automated wafer harvesting" technique that results in better yields, lower labor costs, and also higher efficiency, all of which would help reduce costs per watt.

In more detail, Evergreen Solar has a road map to reduce costs by \$1.65/W to reach \$1.00/W by 2014. Out of this, 65 cents/W is expected to be due to throughput, yield, and other improvements related to its adoption of Quad technology. A 35 cents-per-watt reduction is expected to come from manufacturing in low cost regions, and a further 40 cents from cuts in the cost of materials. Another 25-cent decrease related to yields and efficiency would bring costs down to \$1.00/W. In regions with high grid rates, Evergreen Solar expects to be able to be cost competitive with the grid as early as 2010, while in other regions where grid prices reflect the global average, it expects to reach grid parity by 2012.

The notion of an experience curve is useful when analyzing Evergreen Solar's track record vis-à-vis cost reduction. The term "experience curve" refers to the idea that as production volume increases, cost per watt tends to decrease. It is commonly expressed as the percentage reduction in cost when production volume doubles. Evergreen Solar had achieved an 80 percent experience curve by mid-2008 and expects to drive down costs further. If the company proceeds along the curve as expected, it will be able to cut costs a further 20 percent as megawatt production is doubled (Evergreen Solar Capital Markets Day, June 19, 2008).

SunPower

SunPower also has a goal to reduce installed systems cost by as much as 50 percent by 2012, out of which two-thirds is expected to be achieved as early as 2010. SunPower's preferred method for calculating systems cost is by the levelized-cost-of-electricity method.

As with other companies, SunPower has also attempted to rein in raw material costs by negotiating better terms on its polysilicon contracts. Further, it is moving to thinner wafers (145 microns, from 165 microns) to reduce silicon consumption and is also shifting to larger-sized panels to reduce manufacturing and system integration costs. Since SunPower is forward-integrated into installation, it is specially well-positioned to drive down systems costs. Indeed, SunPower has increased its focus on manufacturing PV systems to the point that they are easier to install, and the company has also been trying to improve installation-related logistics.

First Solar

First Solar is another company that has a clear plan for cost reduction. It already has by far the lowest cost technology for mass production of solar PV, and through its various cost reduction initiatives, intends to drive down systems costs toward grid parity. In terms of ramping up, First Solar has developed the Copy Smart technology, which makes it easier to install additional production lines and expand capacity. Further, the company's strategy of selecting low cost manufacturing locations (Malaysia) may also help drive cost reductions.

Q-Cells and Canadian Solar

In order to reduce the total costs, some companies are using low cost raw materials. There are a few alternatives that can be used as raw material and have costs lower than traditional crystalline silicon.

One such example is Q-Cells' use of upgraded metallurgical silicon. The company has planned a major push into upgraded metallurgical silicon as part of its effort to reduce silicon costs. Q-Cells signed the BSI/Timminco contract for upgraded metallurgical silicon for very high supply volumes of 410 tons in 2008, 3,000 tons in 2009, and 6,000 tons per year from 2010 till 2013. If it transpires as planned, this alternative raw material will lead to a huge cost reduction for Q-Cells.

The other company taking the same route to cost reduction is Canadian Solar. It has also signed a long-term supply agreement with Timminco. The company is expected to purchase up to 5,000 metric tons of upgraded metallurgical silicon through 2011 at a price substantially lower than the current contract prices for polysilicon.

Renewable Energy Corporation

Renewable Energy Corporation has a road map for cost reduction through 2010. The original plan required cutting costs by 50 percent from 2005 levels by 2010. This would involve the following steps (Renewable Energy Corporation presentation at Piper Jaffray conference, June 25, 2008, slide 14):

- Use of FBR technology
- Reducing thickness of wafers

- Quality improvements
- Better cell and module technology
- Constant improvements in manufacturing and process-related areas

LDK Solar

LDK Solar has the following plans for cost reduction (LDK Solar Analyst Day presentation, San Francisco, July 16, 2008):

- Achieving economies of scale
- Reducing wafer thickness
- Backward integration into polysilicon
- Reducing kerf loss by using thinner wires
- Recycling of silicon material
- Improving yields by reducing breakage

In terms of costs, LDK Solar already has one major advantage: being located in low cost China.

Trina Solar

Trina Solar targets the following broad areas in its efforts to cut costs per watt (Trina Solar investor presentation, September 2008):

- Reducing wafer thickness
- Improving conversion efficiency
- Reducing breakage
- Reducing kerf loss

As a result of these efforts, silicon consumption for Trina Solar has decreased from ~8.5 grams per watt at the end of 2006 to ~7.2 grams per watt as of Q208. This decrease in silicon consumption translates into reduced costs per watt for Trina Solar.

EXCHANGE RATE IMPACT

The fluctuations in exchange rate have an impact on the solar PV sector. The impact on a particular company depends on the level of exposure it has to a particular currency. Due to the global nature of the solar PV industry, a company may have its operations in country A and sales in countries B, C, and D. Also, the company may have agreements for major contracts in some other country, E. In this case, any major fluctuation in the currencies of countries A, B, C, D, and E will have an impact on the earnings of the company.

In order to tackle the foreign currency risk (the risk due to the exposure to different currencies), these companies hedge the risks. Some companies enter foreign exchange forward contracts to manage their exposure to foreign currency risks.

A look at some specific company examples will provide a better understanding of how they tackle the foreign currency risk.

Canadian Solar

Canadian Solar has its operations in many countries and thus is involved with multiple currencies. It sells and buys in a wide variety of currencies, particularly in the euro, the U.S. dollar, and the RMB (renminbi). Earlier, a significant amount of the company's sales were denominated in U.S. dollars, but now a majority of the sales are denominated in euros. Also, Canadian Solar has multiyear supply contracts with many suppliers and has made advance payments for specified amounts of silicon wafers. The contract prices are fixed in euro or RMB currency denominations. Major costs and expenses are in RMB, as they are primarily related to domestic sourcing of solar cells, wafers, silicon and other raw materials, toll manufacturing fees, labor costs, and local overhead expenses. The company also has loan arrangements with Chinese commercial banks that are denominated in U.S. dollars and RMB.

Hence, fluctuations in currency exchange rates (particularly among the U.S. dollar, RMB, and euro) could have a significant impact on the company's earnings.

Canadian Solar has both euro-to-dollar and RMB-to-dollar hedging operations, as the company has a reasonable exposure to both the currencies. The company also has plans to expand the hedging operations whenever it is required. A reasonable portion of sales and purchases are done in U.S. dollars, and thus the company is focusing on hedging it further going forward.

As of 2Q08, Canadian Solar had sales operations in many countries: 89 percent in Europe, 6 percent in Asia, and the remainder in

North America. Hence, the company faced the respective currency exposures. The major COGS for Canadian Solar is in RMB. As of June 2008, the company had not entered into any hedging arrangements.

China Sunergy

China Sunergy has more than 90 percent exposure to RMB and the euro because most of its sales are in those currencies. Hence, those two currencies can potentially impact the earnings for the company in case of currency fluctuations.

Energy Conversion Devices

As of Q208, nearly all of ENER's sales and costs were dollar-denominated, so exchange rate impact was minimal.

First Solar

The fluctuations in foreign exchange rates may impact First Solar, as they may affect the product demand and in turn the company's profitability in U.S. dollars to the extent of foreign currency exposure in the price of solar modules and the cost of raw materials, labor, and equipment.

In order to tackle the foreign currency risk, First Solar has foreign exchange forward contracts and other derivatives to hedge its cash flows. The company has about 50 percent hedge on its expected net income through forward contracts over a rolling 12-month period. It has a major percentage of its expected earnings hedged; about 73 percent of the expected earnings going forward are hedged. This simply shows how focused the company is in order to avoid any foreign currency risk.

In the current scenario, most of First Solar's sales are denominated in euros. For 2H08, FSLR has 54 percent of its euro-denominated sales hedged at an average exchange rate of $1.50 per euro. In all, ~73 percent of expected 2008 earnings are hedged. (Note: the company can accomplish a hedge of 73 percent of earnings with just 53 percent of sales hedged because of its natural hedge, that is, the Oder/Frankfurt facility, which allows costs and sales to be matched in denomination.)

LDK

LDK mostly has China-based operations and currently has a 10 percent exposure to the euro. LDK has a very diversified customer base; the revenues come approximately 33 percent from China, 34 percent from Asia Pacific (excluding China), 29 percent from Europe, and 4 percent from North America. So the company faces the exchange rate risks in similar ratio to the currencies of these countries. The company has completed certain corporate hedges in order to manage the foreign currency risk.

Renesola Ltd.

Renesola Ltd. operates mostly in RMB, as most of the company's sales contracts are in China. The company has limited exposure to U.S. dollars, euros, and Japanese yen. Any fluctuations in currency exchange rates between the U.S. dollar and RMB, and between euros and RMB, could have a significant impact on the company's earnings. Hence, the company has employed derivative financial instruments to manage the exposure to fluctuations in foreign currency exchange rates, including forward exchange contracts to hedge the exchange rate risk arising from future costs and capital expenditures.

SunPower

SunPower has foreign currency exposure arising from sales (revenues), capital equipment purchases, prepayments, and customer advances denominated in foreign currencies. In the Components segment, the company has foreign currency exposure from the revenue denominated in euros, including fixed price agreements of SunPower with Conergy and Solon. SunPower has significant percentage of the revenue denominated in U.S. dollars too. The company generates more than 60 percent of its Components segment revenue outside the United States.

On the costs side, a part of the Components segment costs are incurred and paid in euros, and another small part of the Components segment expenses are paid in Philippine pesos and Japanese yen. Also, the company has made prepayments to Wacker-Chemie

AG and has taken customer advances from Solon that are denominated in euros.

SunPower is exposed to the foreign currency risk of a decrease in the value of the euro relative to the U.S. dollar, as this would decrease the company's total revenue. Changes in exchange rates between the U.S. dollar and other foreign currencies may impact the operating margins, because if the foreign currencies appreciate against the U.S. dollar, it would be more expensive for the company to purchase inventory or pay expenses with foreign currencies in terms of U.S. dollars. Also, U.S. dollar devaluation may result in a loss to SunPower, as its products in the United States will be relatively more expensive than products manufactured locally.

On the other hand, an appreciation in the U.S. dollar relative to foreign currencies could make the company's solar cells more expensive for international customers, thus leading to a decline in the company's sales and profitability.

Also, many of SunPower's competitors are foreign companies, and they could benefit from such a currency fluctuation, making it more difficult for SunPower to compete with those companies.

In order to tackle the foreign currency risk, the company conducts hedging activities, using financial tools like currency forward contracts and options. SunPower has a hedging policy that is very calculated, or math-driven, and hence includes no speculation.

As of 2Q08, SunPower had greater than 80 percent of its revenues from Europe. Around 90 percent of the company's euro exposure was hedged. (For the year ending December 30, 2007, about 49 percent of the company's Systems segment revenue was generated outside the United States, out of which 47 percent was denominated in euros. Also, a significant portion of its costs are incurred and paid in euros.)

Suntech Power Holdings

Suntech Power uses RMB as its functional currency due to the PRC government regulations and records transactions denominated in other foreign currencies at the rates of exchange at the time of transaction. The U.S. dollar is the currency used for reporting the financials. A significant portion of the company's total sales are denominated in U.S. dollars and euros, with the remaining portion in RMB,

Japanese yen, and other currencies. On the other hand, a significant portion of the costs are denominated in U.S. dollars, RMB, euros, and Japanese yen. Therefore, any fluctuation in the exchange rates of the U.S. dollar, RMB, euro, and Japanese yen could result in foreign exchange losses and affect company earnings.

Since the financial statements are in U.S. dollars, any appreciation in other currencies against the U.S. dollar will result in an exchange gain, in case the company has assets denominated in the other currencies, and an exchange loss, in case the company has liabilities denominated in the other currencies. For instance, if the Japanese yen appreciates against the U.S. dollar, any new Japanese yen denominated investment or expenditure would become more costly for the company.

Suntech Power manages the foreign currency exchange risk by multiple methods. One way is to increase the sales denominated in euros and then partially lock in the euro versus the U.S. dollar forward contract, using financial tools to hedge against the risk of RMB appreciation. The company thus sources a huge amount of raw materials in the U.S. dollar contract at minimum risk.

Trina Solar

Trina Solar has maximum exposure to U.S. dollars, as it conducts most of its operations in this currency. As of Q208, ~70 percent of TSL's sales were dollar-denominated. The U.S. dollar is used as the functional and reporting currency of the company, although the financial records of certain company subsidiaries are maintained in local currencies other than the U.S. dollar, such as RMB. Trina Solar faces foreign currency risk in this case, since the RMB is not a freely convertible currency.

The PRC State Administration for Foreign Exchange, under the authority of the PRC government, controls the conversion of RMB to foreign currencies. The value of the RMB is subject to changes of central government policies and international economic and political developments affecting supply and demand in the China foreign exchange trading system market. In order to manage this foreign currency risk, Trina Solar holds derivative financial instruments and also enters contracts that contain embedded foreign currency forward contracts.

Yingli Green Energy

Yingli Green Energy has most of its sales in U.S. dollars and euros, and some sales in RMB. On the cost side, a substantial portion is denominated in U.S. dollars, RMB, Japanese yen, and euros. The company makes payments to overseas suppliers in foreign currencies by converting RMB. The change in the value of the RMB against the dollar, euro, and other currencies is affected by changes in China's political and economic conditions, among other things. The PRC government imposes control over its foreign currency reserves in part through direct regulation of the conversion of RMB into foreign exchange and through restrictions on foreign trade.

Hence, foreign currency exchange rates can have a significant impact on Yingli Green's operational results. The company faces foreign currency risk in multiple transactions, such as sales of PV modules in overseas markets, purchases of silicon raw materials and equipment, and the time gap between signing the sales or supply contracts and cash receipts of or disbursements to the contracts. Also, currency risk arises through factors such as the prepayments to suppliers and bank borrowings that will be denominated in a particular currency.

Yingli Green did not use any forward contracts, currency options, or borrowing to hedge the exposure to foreign currency exchange risk until 2Q08. The company plans to use hedging arrangements to reduce the effect of such exposure but faces the hindrance of the limited availability of hedging instruments in China.

As of 2Q08, Yingli Green Energy had most of its sales in U.S. dollars and euros, and some sales in RMB. On the cost side, a substantial portion was denominated in dollars, RMB, Japanese yen, and euros.

JA Solar Holdings Col Ltd.-ADR

JA Solar Holdings conducts most of its operations in China and thus has maximum exposure to RMB. Any fluctuation in the value of RMB against other foreign currencies may have an impact on the company's earnings. In addition, any change in the value of RMB against the euro or other currencies will also have an impact on the company. If the company gets a refund from a supplier for a prepayment made in a foreign currency, it risks foreign exchange losses.

The effect of currency fluctuations can be better explained with the help of an illustration. If the RMB appreciates against the U.S. dollar, it will have an adverse effect on JA Solar, having received the initial public offering in U.S. dollars, when it converts that sum. Its value in RMB would be less. If the RMB appreciates against any other foreign currency, the company's solar cells become more expensive for the international customers and hence reduce its competitiveness, in turn benefiting the foreign competitors.

Evergreen Solar Inc

Evergreen Solar attempts to make all purchases denominated in U.S. dollars, but it also has two multiyear polysilicon supply agreements that are denominated in euros. Also, going forward the company has plans for geographic expansion that will further expose it to many other currencies. Any unfavorable fluctuation in these foreign currency exchange rates may substantially increase the cost of products for the company and have a negative impact on earnings.

Therefore, in order to tackle the foreign currency risk, Evergreen Solar uses derivative financial instruments to manage foreign currency exchange risks.

As of 2Q08, as already mentioned, Evergreen Solar made almost all purchases in U.S. dollars, but it also has two multiyear polysilicon supply agreements that were denominated in euros. The first agreement was with Wacker Chemie AG, for which the company has made a payment of about 9 million euros, and the second agreement was with Silpro, as part of which Evergreen Solar agreed to loan Silpro 30 million euros at an interest rate of 3.0 percent compounded annually.

The company had not entered into any hedging transactions to reduce its exposure to foreign currency exchange risk as of 2Q08.

Solarfun Power Holdings

Solarfun Power operates in China and has exposure mostly in RMB.

NOTE

1. Evergreen Solar annual report 2007, p. 9.

CHAPTER 8

Rebalancing a Solar Portfolio

In an industry that is growing upwards of 50 percent per year in recent times, with the potential to eclipse all other investment opportunities, knowing how to identify winners and losers must be defined by more than whether a stock is "up" or "down" a lot. As the credit crisis puts unimaginable pressure on the solar stocks, the opportunity to invest at historically low valuations exists. During 2008, however, the tremendous excitement that surrounded the solar industry sometimes verged on the "irrational exuberance" of the 1990s. As investor mania reached a pinnacle during the Nasdaq bubble of 1999, informed investors and traders judged it as a time to sell. In the fall of 2001, after 9/11, the bubble burst and tech stocks had sunk to their lows. Portfolio management during this period was crucial and the work of stock selection never more important. We are indeed at a similar inflection point again. While no two periods are ever exactly alike, the credit crisis has ushered in new and healthy appreciation for picking the "real" winners and keeping clear of the losers, stocks that could easily go the way of e-toys, Boo.com and Webvan.

Leveraging the knowledge and principles of earlier chapters, this chapter puts everything into perspective. How does an investor judge when a mania has taken stocks to euphoric levels and decide it's time to sell? How can investors recognize when a terrific company with potential to revolutionize the electric power markets around the globe is on sale? At what point is taking a loss wise? Reacting to changes in the stock market and having the background to anticipate change will be our central focus. Investors will learn how to judge exactly what's "overdone," what constitutes a bonafide opportunity, and how to recognize a mistake before the stock price tells you.

REASONS FOR REBALANCING A SOLAR PORTFOLIO

An investor might want to rebalance a solar portfolio for several reasons. For example, the investor may believe that one or more solar stocks have become overvalued or undervalued relative to the universe of solar stocks. Alternately, the solar stock universe itself may become overvalued or undervalued compared to the broader market. Whatever the reason, an investor would want to have metrics at hand to help decide whether and when to rebalance a portfolio comprising solar stocks.

One of the first steps in rebalancing a solar portfolio is evaluating whether a stock's valuation is out of sync with fundamentals. A simple yet effective way to do this is to compare the key ratios introduced in Chapter 4, such as the P/E (price-to-earnings), PEG (price-to-earnings-growth), P/S (price-to-sales, P/B (price-to-book), EV/EBITDA and EV/sales, within the solar stock universe, as well as against the broader market.

However, even within the solar stock universe there are a variety of companies that may not be directly comparable with one another. For example, MEMC Electronic Materials is at the polysilicon end of the solar value chain. It may not be accurate to compare its relative valuation ratios with, say, Akeena Solar, which is a system integrator/installer. To give another example, First Solar is a company that makes cadmium telluride–based thin film PV. It would not be useful to compare it with companies such as Applied Materials or Centrotherm, which are equipment manufacturers for solar PV.

The idea here is that in comparing companies based on relative valuation, the comparisons must be useful and accurate. To this end, it's necessary to classify companies according to their primary business and position in the value chain. The solar industry can be classified into the following broad segments: polysilicon, cells/ modules, system integration/installation, vertically integrated, and equipment.

Before describing when and how a solar portfolio may be rebalanced, it's helpful to consider a detailed example that illustrates how a "real life" portfolio of solar stocks might appear. This may provide a better understanding of the various factors involved in constructing and rebalancing a solar portfolio. For this purpose, the portfolio of the U.S. Department of Energy is a particularly useful example because the Department of Energy invests in solar companies across the spectrum of technologies in solar photovoltaics and in various stages of maturity ("Venture Investing in Solar— 2007/2008, Investment Theses, Market Gaps and Opportunities," presentation by Craig Cornelius, program manager). While this portfolio may change according to the preferences and requirements of the Department of Energy, it still gives a useful insight into what a solar portfolio might look like. The reader will note that this list includes companies in the solar capital equipment manufacturing business as well:

- Wafer silicon: BP Solar, CaliSolar, Dow Corning, Evergreen Solar, GE, Solaria, Solarworld, Specialized Technology Resources, Spire, SunPower
- Film silicon: Blue Square, USO
- Copper indium gallium selenide (CIGS): Global Solar, Miasole, Nanosolar, SoloPower
- Cadmium telluride: AVA Solar, First Solar, PrimeStar
- Concentrator photovoltaics: Amonix, Boeing (Boeing has a division called Spectrolab, which is a technology leader in solar), EnFocus, MicroLink, SolFocus, Soliant
- Next generation photovoltaics: Konarka, Plextronics
- Nondevice research and development: Dow Chemical, GE, Greenray, Shingleton, Xantrex

The investment profile of Khosla Ventures, a U.S.-based venture capital firm with a focus on the alternative energy industry, gives another example of a well-balanced portfolio of solar stocks.

(The solar portfolio of Khosla Ventures includes companies in the solar thermal space as well.)

- Stion, a company focusing on thin film modules
- Infinia, a company based on the "Stirling generator" concept
- Ausra, a solar thermal company

As one would expect, Khosla Ventures invests in relatively early stage companies that offer the potential for high rewards but come with high risks as well. This is quite different from the approach adopted by the U.S. Department of Energy, which invests in companies in varying stages of maturity instead.

Good Energies, a reputable investor in renewables, has investments in several companies in solar PV, some of which are presented below (Good Energies Web site):

- 6N Silicon: Canadian company that has developed a new technology for solar-grade silicon
- Concentrix: German company that makes concentrator modules and also provides turnkey CPV solutions
- CSG Solar: German company that makes thin film crystalline-silicon-on-glass modules
- Konarka: U.S.-based company that makes organic photovoltaics
- Norsun: Norway-based company that seeks to make low cost monocrystalline ingots and wafers
- Q-Cells: German company based on diverse technologies, including crystalline silicon and multiple thin-film-based technologies
- Solarfun: Chinese company based on crystalline silicon
- Trina Solar: China-based vertically integrated player
- Sunfilm: German company that makes silicon-based tandem junction thin film modules
- SolarReserve: U.S.-based company that develops CSP-based power plants

The bottom line is that Good Energies has invested in a wide variety of solar companies with different technologies and at different stages of development.

Given a basket of solar stocks such as those above, an investor needs to know when and how to rebalance the portfolio. Ratio-based analysis is a good starting point.

In order to compare the key ratios, we will take the mean (simple average) and median (middle value) in each category to avoid any distortions due to outliers. For instance, if a loss-making company is just starting to make profits (a case in point being Energy Conversion Devices as of 1H CY08), its ratios will be distorted and it will be an outlier in the set. In such a case, taking mean as a measure may not be a good idea, so we are also considering the median in the discussion that follows.

RATIO-BASED ANALYSIS

Price-to-Sales Ratio

The price-to-sales or P/S ratio is a simple method to evaluate the value of stocks by taking the market capitalization (the number of shares multiplied by the share price) and dividing it by the company's total sales over the past 12 months. The lower the value of this ratio, the more attractive the investment. One situation where the P/S ratio is particularly useful in determining share valuation is when a company begins to suffer losses and as a result has no earnings. In this case, the price-to-earnings (P/E) ratio cannot be calculated.

Observing P/S ratios of all the solar companies segment by segment during mid-2008 revealed that the mean was highest for cell/module companies at 6.9, followed by vertically integrated companies at 4.54, polysilicon producers at 3.31, equipment companies at 1.66, and system integration/installation companies at 1.5.

The ranking of the segments based on the median P/S value can be seen in Table 8.1. The numbers, not shown in the table, are as follows: vertically integrated companies have the highest P/S median value (3.34), followed by cells/modules companies (3.13), polysilicon companies (1.94), equipment companies (1.33), and finally integration/installation companies (1.14), which have the smallest P/S median value.

In general, cell/module companies and vertically integrated companies both had high mean P/S and median P/S values, as can

TABLE 8.1

Mean and Median P/S values

Mean (Ranks in Descending Order of P/S Value)		Median (Ranks in Descending Order of P/S Value)	
Rank	Segment in Solar Industry	Rank	Segment in Solar Industry
1	Cells/modules companies	1	Vertically integrated companies
2	Vertically integrated companies	2	Cells/modules companies
3	Polysilicon companies	3	Polysilicon companies
4	Equipment companies	4	Equipment companies
5	System integration/installation companies	5	System integration/installation companies

be seen in Table 8.2. In the cell/module segment, First Solar has a very high P/S value of 25.85, and Energy Conversion Devices and Evergreen Solar are not far behind with P/S values of 14.52 and 11.32, respectively. (One important point to note here is that Energy Conversion Devices had negative earnings in 2007 but began to earn a profit in 2008. Hence ENER's high P/S value may naturally fall with any increase in earnings.) Among the vertically integrated companies, Renewable Energy Corp has a high P/S value of 11.1. A risk to placing a premium on REC includes the possibility that its key profit driver, polysilicon, will experience a decrease in earnings as polysilicon supplies expand potentially too fast compared to demand.

Price-to-Earnings Ratio

The price-to-earning ratio (P/E) is a valuation ratio of a company's current share price compared to its per-share earnings. The P/E ratio is calculated by dividing the market value per share by the earnings per share (EPS). P/E is also known as "price multiple" or "earnings multiple," because it denotes how much investors are willing to pay per dollar of earnings. This means that if a company's stock is currently trading at a multiple (P/E) of 10, the investor is willing to pay $20 for $2 in current earnings. A high P/E usually indicates higher earnings growth expected in the future, while a low P/E indicates the opposite.

TABLE 8.2

Price-to-Sales Ratio in the Solar Industry

Polysilicon	P/S	Cells/Modules	P/S	System Integration/Installation	P/S	Vertically Integrated	P/S	Equipment	P/S
Tokuyama Corporation	0.61	Sunways AG	0.31	S.A.G Solarstrom AG	0.42	Canadian Solar Inc	1.39	OC Oerlikon Corp AG	0.75
Sumco Corp	1.18	China Sunergy Co Ltd-ADR	1.45	Phoenix Solar AG	1.14	Trina Solar Ltd-Spon ADR	1.56	PVA Tepla AG	1.06
Wacker Chemie AG	1.49	JA Solar Holdings Co Ltd-ADR	1.67	Akeena Solar Inc	2.95	Solarfun Power Hold-Spon ADR	1.93	BTU International Inc	1.32
Shin-etsu Chemical	1.94	Motech Industries Inc	2.35			Solarworld AG	4.74	Amtech Systems Inc	1.33
DC Chemical Co. ltd	4.45	Solar Enertech Corp	2.41			Sunpower Corp-Class A	6.52	Spire Corp	1.6
MEMC	5.37	Suntech Power Holdings-ADR	3.84			Renewable Energy Corp AS	11.1	Centrotherm Photovoltaics AG	2.62
Timmincio	8.11	Q-Cells AG	5.26					Applied Materials Inc	2.93
		Evergreen Solar Inc	11.32						
		Energy Conversion Devices	14.52						
		First Solar Inc	25.85						
Mean	3.31	Mean	6.90	Mean	1.50	Mean	4.54	Mean	1.66
Median	1.94	Median	3.13	Median	1.14	Median	3.34	Median	1.33

Source: Bloomberg

T A B L E 8.3

Mean and Median P/E Values

Mean (Ranks in Descending Order of P/E Value)		Median (Ranks in Descending Order of P/E Value)	
Rank	Segment in Solar Industry	Rank	Segment in Solar Industry
1	Cells/modules companies	1	Cells/modules companies
2	Vertically integrated companies	2	Equipment companies
3	Equipment companies	3	Vertically integrated companies
4	Polysilicon companies	4	System integration/installation companies
5	System integration/installation companies	5	Polysilicon companies

Tables 8.3 and 8.4 show the ranking of solar company segments based on P/E analysis conducted in mid-2008. The actual P/E mean values in Table 8.4 show cell/module companies at 70.68, followed by vertically integrated companies (44.4), equipment companies (32.71), polysilicon companies (17.62), and then system integration/installation companies (17.3).

Ranking the segments based on the median P/E value (Table 8.3), cell/module companies have the highest P/E median value (59.76), followed by equipment companies (30.86), vertically integrated companies (26.7), system integration/installation companies (17.3), and then polysilicon companies (13.75).

In general, companies at the polysilicon production level of the solar PV life cycle have low P/E values, while companies manufacturing cells and modules have higher P/E values.

P/E Value: Is It a Good Comparison Ratio?

Energy Conversion Devices is a good candidate to test the validity of the P/E comparison ratio, as the company has just recovered from losses and has fast growing earnings. In such a case, P/E value might not be a good comparison ratio. Energy Conversion Devices had been incurring losses and just recovered from them. This raised its P/E value tremendously. In this instance, P/E value will not give the correct picture to a portfolio manager or an investor.

TABLE 8.4

Price-to-Earnings Ratio in the Solar Industry

Polysilicon	P/E	Cells/Modules	P/E	System Integration/Installation	P/E	Vertically Integrated	P/E	Equipment	P/E
Sumco Corp	7.47	Motech Industries Inc	15.01	S.A.G Solarstrom AG	14.05	Trina Solar Ltd-Spon ADR	14.73	PVA Tepla AG	17.50
Tokuyama Corporation	9.92	Suntech Power Holdings-ADR	32.08	Phoenix Solar AG	20.55	Canadian Solar Inc	22.28	Applied Materials Inc	19.53
Wacker Chemie AG	12.96	Q-Cells AG	47.85			Solarfun Power Hold-Spon ADR	25.48	Amtech Systems Inc	30.86
Shin-etsu Chemical	14.53	Sunways AG	71.67			Solarworld AG	27.91	Centrotherm Photovoltaics AG	32.21
MEMC	16.89	First Solar Inc	101.32			Renewable Energy Corp AS	56.31	BTU International Inc	63.44
DC Chemical Co. Ltd	43.97	Evergreen Solar Inc	156.17			Sunpower Corp-Class A	119.66		
		Energy Conversion Devices	3,800.00						
Mean	17.62	Mean	70.68	Mean	17.30	Mean	44.40	Mean	32.71
Median	13.75	Median	59.76	Median	17.30	Median	26.70	Median	30.86

Source: Bloomberg

Price-to-Book Value Ratio

The price-to-book value (P/B) or price-equity ratio is used to compare a stock's market value to its book value. Book value is an accounting term that denotes the portion of the company held by shareholders.

There are two methods to calculate P/B values. The first involves dividing the company's market capitalization by the company's total book value from its balance sheet. The second method uses per-share values and involves dividing the company's current share price by the book value per share (that is, its book value divided by the number of outstanding shares). A higher P/B ratio implies that investors expect the company management to create more value from a given set of assets and/or that the market value of the firm's assets is significantly higher than the accounting value. A low P/B ratio could mean that the stock is undervalued. However, it could also mean that something is fundamentally wrong with the company. This ratio gives investors an idea whether they are paying a higher value than the value of the company if it went bankrupt immediately.

Using P/B as a valuation ratio has some advantages. It provides a relatively stable and intuitive measure of value that can be compared to the market price. It can be compared across similar firms for under- or overvaluation (as there are reasonably consistent accounting standards across firms). And even firms with negative earnings that cannot be valued using P/E ratios can be evaluated using P/B ratios.

Ranking the solar company segments on the basis on P/B ratios during the mid-2008 period revealed that cell/module companies have the highest P/B mean value (6.04), followed by polysilicon companies (4.64), vertically integrated companies (4.53), equipment companies (4.07), and finally system integration/installation companies (3.14).

Ranking the segments based on the median P/B value revealed that the cell/module companies had the highest P/B median value (4.02), followed by vertically integrated companies (3.43), polysilicon companies (3.32), equipment companies (3.25), and finally system integration/installation companies (2.78).

Gross Margins

Gross margins provide another comparison basis for solar PV companies. Again we conducted an analysis during the mid-2008 period ranking the segments based on mean gross margin values. The analysis revealed that equipment companies had the highest gross margin mean value (30.75), followed by polysilicon companies (29.95), cell/module companies (23.61), system integration/installation companies (21.23), and finally vertically integrated companies (17.9).

Ranking the segments based on the median gross margin value revealed that polysilicon companies had the highest gross margin median value (31.20), followed by equipment companies (26.50), cell/module companies (21.70), system integration/installation companies (21.23), and finally vertically integrated companies (19.08).

In general, the gross margin decreases as we move further down the value chain—polysilicon companies that are at the upstream end of the value chain had the highest gross margins, followed by the cell/module companies, and finally the system integration/installation companies. To cite an example, MEMC, a company in the polysilicon segment, had high gross margin of about 52 percent because of the tight poly supply. Vertically integrated companies have the least gross margin. As or when the credit crisis abates, the investor must reassess the market and determine where profitability has migrated across the value chain and which segments will have the best profit opportunities. Estimating these tectonic shifts in value chain economics will command intense efforts. Chapter 9 will address what appears to be changing and can focus the savvy mind on what may be important. However, before we skip ahead and potentially miss more of the critical analysis steps needed to evaluate the stocks, let's continue on and look more closely at earnings.

Earnings Growth

Earnings growth is the measure of growth in a company's net income over a specific period, usually one year. It can be calculated using the actual data from previous periods or estimated data for future periods.

Mean earnings growth during the analysis period (mid-2008) was highest for the cell/module companies (more than 300 percent), followed by polysilicon companies (~57 percent), equipment companies (~19 percent), vertically integrated companies (~7 percent), and finally system integration/installation companies (~2 percent).

The median earnings growth followed a somewhat similar trend, only with the exception of polysilicon companies having the maximum median earnings growth of ~50 percent, followed by cell/module companies (~29 percent), and equipment companies (~12.5 percent). Vertically integrated and system integrated/installation companies have negative median earnings growth of –24 percent and –49 percent, respectively.

Price/Earnings-to-Growth Ratio

The price/earnings-to-growth ratio (PEG) is used to determine a stock's potential value taking into account the company's earnings growth. It is calculated by dividing the price/earnings ratio by the annual long-term earnings per share (EPS) growth of the company.

PEG is a better valuation ratio than the P/E ratio because it also accounts for growth. Similar to the P/E ratio, a lower PEG means that the stock is more undervalued. The drawback to PEG is that projected numbers are used, and hence the resultant figure can be less accurate.

COMPANY ILLUSTRATIONS

A few specific company examples will provide a better understanding of the valuation figures.

Energy Conversion Devices

Among cell/module companies, Energy Conversion Devices was an outlier when it comes to the P/E value. Its P/E was extremely high at 3,800 because the company was incurring losses for many years and only recently started making some profit. So for our analysis we have excluded the company Energy Conversion Devices, given its special set of circumstances and because it is clearly an outlier in the set.

According to the valuations given by *Forbes,* the estimated P/E for Energy Conversion Devices was 45, 23.9, and 16.2, for 2008, 2009, and 2010, respectively. Also, the PEG ratio for Energy Conversion Devices was 1.0. The *Forbes* consensus revenue figures during the analysis period were $468 million and $673 million for 2009 and 2010, respectively (http://finapps.forbes.com).

First Solar

First Solar's P/E ratio during the analysis period was extremely high, 101.3. The company had high returns of investment in 2007 and was also displaying strong earnings estimates in 2008. It had a trailing P/E of 136. According to the valuations given by *Forbes,* the estimated P/E for First Solar at the time of our analysis was 74.5, 39.7, and 27.4, for 2008, 2009, and 2010, respectively. Also, the PEG ratio for First Solar was 0.9. The *Forbes* consensus revenue figures were $1,228 and $2,200 million for 2008 and 2009, respectively (http://finapps.forbes.com/finapps/jsp/finance/compinfo/Earnings.jsp?tkr=FSLR).

The most important differentiating factor for First Solar then and now is that it is largely immune to the polysilicon shortage, as it is a larger player within the thin film PV space. This new technology being used by the company is not as efficient as crystalline silicon in converting sunlight energy to electricity, but the low cost of manufacturing the product is one of the primary selling points. Also, thin film PV has the ability to work under lower light conditions. With the much expected polysilicon shortage in the near future, First Solar has a huge advantage, which explains the strong estimated consensus valuations for the company. Analyzing First Solar will require estimating the company's growth potential, earnings power, and the degree to which the value proposition and business model can remain superior to other investment opportunities. Even during the credit crisis investors have been willing to pay up for FSLR shares. This, however, may not always be true especially if environmental concerns, business plan changes, or rare raw materials become less available changing the picture for the company going forward. The lesson here is to understand the investment risks of even the strongest companies and not to merely adopt the market's view of the best or worst. To follow the crowd more often than not obfuscates both risk and reward.

SunPower

Among vertically integrated companies analyzed during mid-2008, SunPower had a high P/E value of ~120, compared to the mean segment P/E value of 44. According to the *Forbes* valuations, the estimated P/E for SunPower was 42.1, 26.8, and 20.8, for 2008, 2009, and 2010, respectively. Also, the PEG ratio for SunPower was 0.8. The *Forbes* consensus revenue figures were $1,421 and $ 2,053 million for 2008 and 2009, respectively (http://finapps.forbes.com).

During the analysis period the company's strategy of vertical integration was expected to help it combat the shakeout, since SunPower would be in a position to capture a larger area in the solar PV chain and have more control of the module sales distribution channel. SunPower's major differentiating factor is its focus on higher cost, higher efficiency, and higher-quality solar cells. The company has recently acquired PowerLight and has captured a huge market for high-quality commercial and industrial installations of solar PV systems. However, despite vastly improved control over its destiny given its integration strategy, the company has not experienced a protracted downturn in its current form. The stresses of module and systems price declines in the market could expose weaknesses in the company's margin stacking strategy, leaving it with excess margin pressure should competing module supplies become vastly cheaper than in-house production. As competitors close the gap in efficiency with SunPower using less expensive materials and processes, SunPower may become competitively disadvantaged.

Currently, SunPower has a conversion efficiency of 23.4 percent, which is much higher than the industry average (http://www .renewableenergyaccess.com/rea/news/story?id=46286).

To get a better idea of the industry conversion efficiencies, let's take a look at this percentage for some other top companies:

- Suntech Power Holdings (polysilicon): 18 percent (Suntech product brochure)
- Sharp (polysilicon): 13 percent (SharpUSA Product Brochure)
- Kyocera (polysilicon): 18.5 percent (Kyocera, Solar Timeline)
- Solarfun (polysilicon): 17.2 percent (Solarfun Web site)
- JA Solar Holdings (monosilicon): 17.7 percent (http:// jasolar.com/Web/products-en.asp)

- Trina Solar (mono- and polysilicon): 16.6 percent (TSL 20-F 2007, 7)
- Evergreen Solar (string ribbon): 15 percent (ESLR 2007 Earnings Call Transcript, 1)
- Energy Conversion Devices (amorphous silicon thin film): 8.5 percent (ENER F1Q08 Earnings Call Transcript, 5)
- First Solar (CdTe thin film): 10.5 percent (http://www.renewableenergyaccess.com/rea/news/story?id=46286)
- DayStar Technologies (CIGS thin film): 14 percent (DSTI 10-k 2007, 1)
- Ascent Solar (CIGS thin film): 9.5 percent (Ascent First Quarter 2008 10-Q, 23)

BTU International Inc.

Among the equipment companies, BTU International had a high P/E value of 63.44, compared to the mean value of 32.71. According to the valuations given by *Forbes*, its estimated P/E was 75.3, 17.2, and 32.6, for 2008, 2009, and 2010, respectively. The *Forbes* consensus revenue figures were $77 and $97 million for 2008 and 2009, respectively (http://finapps.forbes.com).

DC Chemicals

DC Chemicals' P/E value was 43.97, compared to the mean P/E value of 17.62 for the polysilicon companies segment.

DIFFERENTIATED COMPANIES IN THE SOLAR PV SECTOR

High market valuations enjoyed by certain companies in the solar PV sector as measured by P/S, P/E, and P/B generally reflect significant improvements in their financial performance during the analysis period.

Based on the case analysis earlier in this section, the two most visible solar PV companies are First Solar (FSLR) and SunPower (SPWR). FSLR with a P/S value of 25.85, compared to the mean P/S value of 6.9 in the cell/module segment; and SunPower with a P/S value of 6.52, compared to the mean P/S value of 4.54 in the vertically integrated segment. Also, FSLR with a P/E value of 101.32, compared to the mean P/E value of 70.68 in the cell/module segment; and SunPower has a P/E value of 119.66, compared to the mean P/E value of 44.4 in the vertically integrated segment. Another point to be highlighted here is that the vertically integrated companies have the advantage of being positioned to capture the value in the entire solar PV chain.

This strong performance not only ensures the presence of these companies in the solar PV portfolio at the time of the analysis, but also improves the overall visibility and viability of the solar PV business model potentially going forward, and thus may attract more investors, and also funding for solar PV start-up companies.

Comparison with Technology Stocks

To get a better picture of the market valuation of these solar PV stocks, we can compare them with leading technology companies like Google (GOOG) and Cisco (CSCO). Currently, many solar PV companies during the analysis period (mid-2008) trade at higher valuation multiples than GOOG and CSCO. As we have already seen, both FSLR and SPWR trade at premium price-to-sales (P/S) and price/earnings-to-growth (PEG) multiples in comparison to some leading technology stocks. The comparison to the tech stocks is significant for two reasons—first, that the technology sector is relatively new, and second, that the solar PV sector is also dependent on new technologies as a prime growth factor.

The PEG ratios at which FSLR and SPWR trade are about 2.9 times and 2.2 times, respectively, while CSCO and GOOG trade at about 1.3 times and 1.2 times, respectively. In comparison, the S&P 500 index trades at a PEG of 0.8 times, while the technology and energy segments trade at about 0.6 and 2.2 times, respectively.

Comparison to a Similar Sector

It is useful to have a reference point against which we can compare the valuations of solar stocks. In this context, the wind energy sector serves this purpose for an investor. Like solar, wind is also a fast-growing industry in the alternative energy space. One can look at the relative valuation ratios in this industry as an indicative benchmark for solar company valuation.

The historical relative valuation numbers for a few companies in the wind energy industry, according to data from Bloomberg (as of September 9, 2008) looks like this:

- Gamesa: P/E 20.39, P/S 2.11
- Nordex: P/E 33.03, P/S 1.78
- Suzlon: P/E 33.68, P/S 2.53
- Vestas: P/E 47.95, P/S 3.19

We can also compare the relative valuations of solar PV companies with those in the semiconductor industry. When it first started out, the semiconductor industry had several characteristics similar to the current state of the solar PV industry. In particular, the industry was (and still is) technology intensive, had high growth rates, and was based on a model where semiconductor companies tended to own their own fabrication units (fabs). As the industry evolved, however, growth rates inevitably slowed, while several companies shifted toward a less asset intensive "fabless" model. This allowed them to take advantage of lower capital and manufacturing costs in countries like Taiwan and China, freed up the balance sheet, and also allowed them to focus on their core technology-related IP.

For these reasons, valuations in the semiconductor industry might offer useful insights into how valuations in the solar PV industry might progress in the long run. Note that the valuation ratios listed below are as of September 15, 2008 (source: Bloomberg), and change as prices and earnings vary; however, these numbers do provide an indication to the investor about "typical" valuations in the semiconductor industry. Also, for companies with negative earnings, only the P/S ratio applies.

- Intel: P/E 15.5, P/S 2.9
- Advanced Micro Devices: P/S 0.54
- Texas Instruments: P/E 11.48, P/S 2.18
- Infineon: P/S 0.79

While these numbers cannot serve as perfect reference points, they can help an investor gauge whether companies in the solar sector are more or less in line with market valuations for companies in a similar industry. On the other hand, if a company's market price gives it a very high P/E or P/S relative to reference numbers, an investor could immediately proceed to evaluate whether these premiums are justified by anticipated growth rates.

PRINCIPLES FOR MANAGING A SOLAR PORTFOLIO

Here are four potential strategies or principles that could be used to manage a portfolio of solar stocks, depending on an investor's requirements.

Top-Down Approach

There are several ways in which a top-down strategy can be used for investing in solar stocks. For example, an investor may decide to allocate certain fixed percentages of her solar stock portfolio to companies in different positions in the value chain, and invest 10 percent in polysilicon companies, 10 percent in wafer companies, 50 percent in cell/module companies, 10 percent in system integrators/installers, and 20 percent in solar equipment companies. In this example, the investor is overweight on cell/module companies as compared to other parts of the value chain. The principle behind such a strategy would be to allocate capital based on the investor's expectations of how different parts of the value chain are expected to perform relative to others.

As the investor detects shifts in the industry structure so value gets transferred from one portion of the value chain to another, she may decide to revisit the top-down portfolio allocation.

Consider again the example of an investor who believes that cell/module companies are poised to capture the maximum value and who has 10 percent in polysilicon companies, 10 percent in wafer manufacturers, 50 percent in cell/module manufacturers, 10 percent in system integrators, and 20 percent in solar capital equipment companies.

If this investor believes at a later point in time more of the value in the future will be captured by polysilicon companies and less by cell/module companies, she may decide to rebalance her portfolio accordingly. This particular shift in the industry could happen for several reasons.

Currently, as we have noted, there are companies with limited or no experience in polysilicon manufacturing that have massive capacity additions plans. It is possible that some of these planned additions may not come online for various reasons. If this happens, the industry may witness continued tight supply of polysilicon, given the large scale capacity expansion plans of cell/module manufacturers. In our example, if the investor senses a shift in where value is created in the solar chain, she might decide to invest, say, 20 percent in polysilicon companies and 40 percent in cell/module manufacturers, with the rest of the allocation remaining unchanged.

Again, after such a high-level allocation, the investor could choose to select stocks based on different criteria. For example, she might choose to simply select the top 5 or 10 companies (in terms of, say, revenue or market capitalization) in each part of the value chain based on her value-chain-level allocation. Or an investor might choose the 5 or 10 stocks in each part of the value chain that have the lowest P/E, P/S, or P/B.

Another top-down portfolio strategy could be to allocate capital based on technology. For example, an investor might decide to allocate 50 percent of her solar portfolio to crystalline silicon companies and 50 percent to thin film companies; within thin film, the allocation might be 25 percent (of the overall solar portfolio) in amorphous silicon-based PV, 50 percent in cadmium telluride, and 25 percent in copper indium gallium selenide (CIGS). The investor's exact percentage allocation would depend on her beliefs in the prospects of each type of technology.

Another strategy would be to allocate capital in the solar sector based on the maturity of the companies and their technologies. The principle for this is based on the premise that a more mature company might be more stable but offer relatively lower growth rates. On the other hand, a company that is comparatively new and less mature might pose greater risks to the investor, but if the company and its technology emerge a winner, the investor stands to make impressive gains. As we saw in our earlier description on the solar portfolio held by the U.S. Department of Defense, for instance, the investments covered a wide range of companies in different stages of maturity and technologies and with different levels of risk. To give another example, a venture capitalist looking for potentially high returns might want to invest mainly in early stage companies that are relatively less mature (and also pose higher risks).

A third top-down portfolio strategy could be to decide allocation geographically, based on the markets of the solar PV companies. For example, an investor could allocate capital so that 50 percent of total revenue (of companies in the portfolio) comes from Germany, 30 percent from Spain, and 20 percent from other countries. The exact allocation would depend on the investor's view on the prospects of these end-markets.

An investor may want to rebalance a top-down solar portfolio under several situations. If an investor with a value-chain-based allocation believes in the prospects of a structural change in the industry so that more of the value gets captured by, say, cell/module companies, she might want to shift her allocation accordingly. Or, if an investor with a technology-based allocation sees an industry shift in favor of thin film, she might want to change her strategy to take this into account.

Bottom-Up Approach

Investors could also adopt a bottom-up approach to the management of a solar stock portfolio. A pure bottom-up strategy would involve studying and analyzing the characteristics of individual companies and deciding purely on the basis of individual company merits whether to invest in the company. Of course, this also means a lot of research effort would be required to analyze and pick only those companies that exhibit the desired characteristics in terms of fundamentals, earnings and revenue growth potential, competitive positioning, and so on.

However, since the universe of solar stocks is very large, it might be necessary to screen stocks based on certain parameters. This preliminary screening can be done in several ways: on the basis of market capitalization or revenues or net profits, to name just a few. Once the preliminary screening is complete, an investor can use the techniques described in Chapter 4 and Chapter 7 (where we describe valuation and investing for the long run, respectively) to decide on specific stocks and the principles described in Chapter 6 to help decide when to pull the trigger and actually buy a stock.

When considering a company such as Q-Cells, for instance, an investor might want to look at its valuation by various methods—discounted cash flow, price-to-earnings, price-to-sales, price-to-book—vis-à-vis peer companies, to consider whether the stock looks overvalued, undervalued, or fair. Drilling down into the company's business and operations, the investor would discover that, in terms of investing for the long run, Q-Cells represents technology diversification.

The company has various types of technologies based on crystalline silicon, including polycrystalline silicon, monocrystalline silicon, string ribbon technology, and low concentration PV technology. Q-Cells has a diverse set of thin film technologies as well, including cadmium-telluride-based thin film, micromorph, copper indium gallium selenide, and crystalline silicon-on-glass (CSG). These are based on glass substrates. Further, the company has flexible substrate-based thin film with amorphous-silicon-on-plastic-foil technology (flexcell), and Q-Cells is also moving into upgraded metallurgical silicon.

Apart from this, our investor would look at management-related commentary, the company's capacity expansion plans, and its track record in terms of execution as part of the bottom-up approach, before deciding whether to invest in the stock.

The primary advantage with such an approach is that the investor would be analyzing all the key aspects of the company in great detail and that the investment would be rooted in the fundamentals of the company. Again, however, this approach would also require a significant amount of effort in determining which companies to pick.

Contrarian Strategy

Investors could also adopt a "contrarian" strategy where they invest in stocks or sectors that are neglected by the broader market. If a stock or subsector (a position in the solar supply chain) has been consistently depressed in the market without strong fundamental reasons for backing, an investor may decide, in line with her contrarian strategy, to buy into the stock or subsector.

For example, if a company has a controversial technology that is not widely accepted by investors, the company's stock may be down. If an investor believes that the technology is indeed viable, she may buy the stock. Upgraded metallurgical silicon (UMG-Si) is one such technology that has been quite controversial in the past: investors following a contra strategy may buy stocks of upgraded metallurgical silicon manufacturers and PV companies that use upgraded metallurgical silicon feedstock if their prices are depressed and if the investors believe the low valuations are not justified.

To give an example of a contra strategy at the subsector level, if the broader market believes that a subsector is not going to fare well and the investor believes that the sector's low valuation is not in line with its fundamentals, she may decide to buy into the subsector. Of course, this strategy involves a lot of research and analysis because an investor would need sound reasons to go against the general opinion of the market.

Market Capitalization-Based Strategy

An investor may choose to create a solar portfolio based on the market capitalizations of stocks. For example, she may choose a large-cap basket of solar stocks that comprises stocks like First Solar and SunPower. Generally (though not always), these stocks would have reasonably well-established technologies, a record of execution, and a pipeline of sales contracts to back their valuations. While these companies offer a more reliable and less risky route to play the solar sector, they might not offer the same investment potential as some of the smaller, riskier, fast-growing stocks. These stocks are also typically much more liquid than the solar sector small caps.

Another strategy for the solar investor is to choose to invest only in smaller companies with market capitalizations less than a certain value. These companies might be based on emerging technologies and may offer the potential for higher growth, though they might come with much higher volatility. They might be cash flow negative and have a cash burn every few months, sometimes just a few weeks, meaning the companies might face liquidity-related risks besides business risks. Also, stocks of these companies could be much less liquid than those with large market capitalizations, further dampening their attractiveness to an investor. In summary, an investor would choose a small-cap strategy if she does not mind bearing the risks we have mentioned, in exchange for potentially higher returns.

CHAPTER 9

The Solar Storm

It was the best of times, it was the worst of times; it was the age of
wisdom, it was the age of foolishness; it was the epoch of belief, it
was the epoch of incredulity; it was the season of Light, it was the
season of Darkness; it was the spring of hope, it was the winter
of despair; we had everything before us, we had nothing before
us; we were all going directly to Heaven, we were all going the
other way.

—*Charles Dickens*

The year 2008 began with storm clouds on the horizon that envel-
oped the industry before the year ended. Once high-fliers, First
Solar (ticker, FSLR), SunPower (SPWR), Q-Cells (QCE GR), and
Suntech Power (STP) were reduced to a fraction of their values at
the beginning of the year. Prices for solar modules fell an astonish-
ing 30 percent. As we began 2008, companies seemingly loved by
the investment community—LDK Solar Co. Ltd. (LDK), Canadian
Solar (CSIQ), and JA Solar (JASO)—were sold in unprecedented
fashion. Investors, overwhelmed by the ferocity of the credit crisis,
began to recognize that the solar market had invested unwisely,
had overbuilt capacities, and in a number of cases may not have a
business model that can sustain their debt. What went wrong, and
what lies ahead for the investor in solar stocks?

As 2008 began, the solar stocks had come off a breathtaking bull run in the fourth quarter of 2007, largely driven by the fact that hedge funds, registered investment managers, and mutual funds—all of which were underweighted in alternative energy—envisioned 2008 as likely to usher in record energy prices. They could not report to their clients that they did not have an investment thesis about energy technology that would benefit from looming high energy costs and was the only substitute that could resolve the crisis in the future. Although the run-up leading into the end of 2007 met with a bust beginning in 2008, most solar stocks enjoyed huge gains after the new year sell-off ended but before the market for global subsidies began to wax and wane during the late summer. This began the season of despair for the uninitiated solar stock investor, who was now outflanked by the well-informed.

The trouble for solar stocks in 2008 began with the failure of the Energy Independence and Security Act of 2007 to include benefits for the solar energy industry. The long-awaited extension of federal tax credits for renewable power generation, scheduled to expire at the end of 2008, was again postponed. The best three-word summary of the final bill, passed on December 19, 2007, is "pork for corn," because of all renewable energy sources covered under the law, none fared as well as ethanol. Solar was more or less completely left out. Less than one year later, VeraSun, the nation's second largest ethanol producer behind Archer Daniel Midland Corporation, declared bankruptcy. It was not until the passage of the "Troubled Asset Relief Program" (TARP), officially called the Emergency Economic Stabilization Act of 2008, that the goals of the solar PV industry in the United States were realized and the Federal Investment Tax Credit was passed into law, a year late and in the throes of an economic downturn often compared in severity with the Great Depression.

The lesson, of course, is that while government subsidies can create opportunities, they often cannot abolish the rules of value creation. The further lesson, for solar energy investors, is to remain acutely attuned to the subsidies for the solar industry, since the industry is not likely to achieve grid parity across a wide enough geography, with the right mix of high electricity prices and high sunlight, for the next several years or probably more.

Since the solar industry remains dependent on subsidies, the lack of additional support from the U.S. federal government exiting in 2007, and the possibility of less support coming from Germany in 2008, painted a grim picture for solar stocks from the beginning of 2008. From January 1, 2008, to March 31, 2008, the solar group of stocks fell precipitously as certain savvy institutional investors fled. For the investor who was not aware of the changes in the offing at home and abroad, the pain of holding volatile solar stock likely became too much. The uninitiated U.S. investor remained mired in the California Solar subsidy program and lost sight of the big picture. It is absolutely critical to solar stock investing not to lose sight of the big picture and to remember at all times that the industry remains dependent on public policy until grid parity is reached in markets sufficiently endowed with enough demand to sustain the industry.

As 2008 began, investors wondered if the California Solar Initiative would result in a boom for solar in the United States and would relieve the pressure. However, before January ended, the signs failed to produce optimism. Only four megawatts of new solar projects were approved in the first half of that month. The implied run-rate of eight megawatts of approvals per month represented the slowest monthly rate since the program's inception. For perspective, project approvals in the second half of 2007 were running at 13 to 15 MW per month, and investors had hoped for evidence that a large portion of the total program's volume—3,000 MW—could come into the market. The evidence became clear that this would not be the case, and the United States would not surprise to the upside for 2008. The situation in Spain, however, was completely different.

As the California solar market fell shy of the expectations of many "solar bulls," investors turned their attention to Germany and Spain, the markets they believed would determine the solar industry's growth rate in 2008. Although the industry was experiencing phenomenal demand in Spain, the poorly conceived Spanish subsidy that resulted in a "solar gold rush" ultimately was too costly and inevitably caused a reversal.

Despite what would turn out to be a record year for the market in Spain, the prospects of a potentially negative revision to the all-important German market was a concern. Throughout the first quarter of 2008, clouds continued to gather, and the solar stock performance in the quarter demonstrated the risk (see Table 9.1). This, however, was not the end of the story. Although solar companies

TABLE 9.1

Selected Solar Stock Performance, 1Q 2008

Company	Performance Rate
First Solar	14%
SunPower Corporation	43%
QCells	35%
Yingli Green Energy	56%

achieved all-time process improvements and greater and greater scale, the investment climate worsened sharply as savvy investors discounted changes to subsidies and a growing case for oversupply. Despite a continuation of stellar quarterly results in both the first and second calendar quarters, investors and traders alike took note that the stocks failed to achieve new highs and increasingly were setting new lows. Trading this volatility with a background in the group became the dominant short-term strategy.

Investors began to sell off stocks as they contemplated a potential reversal after several years of growing momentum for subsidy programs around the world. By February, indications from the stakeholders and policy makers in the Germany market were that they wanted draconian cuts in the subsidy program. Conservatives in the German Parliament began to speak openly about capping the uncapped subsidy program and cutting subsidies by 30 percent.

Such a turn of events could have set back the solar industry for years, slowing growth and reducing research and development. As events unfolded, however, volatility in the German market began to slow, partly as a result of Spain gradually accelerating its buildout. By March 2008 the first bottom was temporarily in for the solar stocks, and traders began to stake out positions for what would ultimately be excellent short-term strategies on an ascendant Spanish market, rising module prices, and expanding margins. The bottom lasted only a month, but the principles used to establish the position are critical to recognize.

Before pulling the trigger on trades in March 2008, professional investors in solar had conducted an analysis of the market on a one- to two-quarter strategy, beginning with a check of the market for supply and demand of solar modules.

The prospect of falling supplies for polysilicon and dramatically increasing demand coming from Spain became the basis for the long play into spring, as the significant secular risk of oversupply and the continuation of the wholesale revaluation of the group failed to deter the market. Professional investors positioned long, but some understood that the revaluation of the group on a secular and cyclical basis would limit rallies for the rest of the year. Those concentrating on the emerging oversupply of polysilicon viewed polysilicon companies MEMC, REC, and Wacker with an eye to selling them in the second half. These professional investors did not look for a significant rally among those companies. Professional investors resorted to modeling supply using a model as shown in Table 9.2.

These savvy investors then modified their view of the demand picture to account for the growing market in Spain (Table 9.3),

T A B L E 9.2

Polysilicon Supply-Driven Model*

Polysilicon Sources and Uses	2007	2008	2009
End-year polysilicon capacity (base-case production estimates)	**55,630**	**81,320**	**127,620**
Current year average capacity	47,285	68,475	104,470
Plant utilization	109%	108%	105%
Annual polysilicon production (base case)	**51,400**	**73,850**	**109,250**
Average year polysilicon production	**44,975**	**62,625**	**91,550**
Less: microelectronics demand	(19,239)	(22,814)	(26,882)
Plus: reject/reprocessed IC silicon	4,281	5,077	5,982
Plus: reject/reprocessed silicon	2,895	4,593	7,413
Available for PV production (in tons)	**32,912**	**49,481**	**78,063**
c-Si PV production potential (in MW)	**3,379**	**5,493**	**8,933**
Thin film PV production	306	681	1,117
Total potential PV production	**3,685**	**6,173**	**10,050**
Key assumptions			
Growth in microelectronics demand	13.4%	13.4%	13.4%
Reclaimed silicon as % of poly-si output	9.0%	9.0%	9.0%
c-Si PV as % of global PV market	92%	89%	89%
Tons of silicon/MW of PV (adj.)	9.7	9.0	8.7
Y/Y improvement in silicon efficiency	11.0%	7.5%	3.0%

*All figures in tons except where otherwise noted.
  Source: AEI estimates, company reports

T A B L E 9.3

AEI 2008 Country Demand Estimates*

Country	Original (Fall 2007)	Revised (February 2008)	Estimate (Change between Periods)
Germany	1.5	1.7	13.33%
Spain	0.35	0.9	157.14%
Italy	0.1	0.3	200.00%
Greece	0.2	0.2	0.00%
Korea	0.15	0.2	33.33%
Japan	0.35	0.35	0.00%
United States	0.3	0.3	0.00%
ROW	0.34	0.6	76.47%
Inventory	0.1	0.3	200.00%
Total	**3.29**	**4.55**	**38.20%**

*Estimates in GW unless otherwise noted.
  Source: AEI estimates

concluded that the oversupply call would be trumped by a trading call, and got long, while not climbing onboard the polysilicon stocks. This turned out to be the right move, as the stocks moved up strongly in March, with many names hitting their highs for the year.

> Polysilicon and c-Si device companies face an increasingly tough supply-demand picture beginning in the second half of the year. We reiterate our oversupply thesis. That said, we HAVE NOT seen any evidence of excessive asp declines, margin pressure or first-hand channel inventory glut to say "we got it"—the smoking gun to validate our fundamental thesis with objective real-time data. Spain is coming on strong pushing up module asps. *Our trading call, therefore, is completely contrary to our fundamental thesis.* We think the solar group will consolidate until shortly after Q-Cells and Suntech Power report next week before moving-up after nearly eight weeks of market capitalization destruction. Indeed, from a purely trading perspective, we think that as the reporting period exhausts itself the group can rally again. We are not sure how long a rally might last; however, with higher oil and solar markets continuing to fire on all cylinders, growth investors could very well return to the group given the degree to which valuation has come down.

> —*AEI to clients, February 2008*

T A B L E 9.4

PV Module Supply

	2007	2008
c-Si PV production potential (in MW)	3,379	5,493
Thin film PV production	306	681
Total potential PV production	3,685	6,173
Demand (in MW)		
Worldwide	3,290	4,500
Supply-demand balance (in MW)	+395	+1,673
Source: AEI estimates, company reports		

Therefore, despite the long trade, professional investors knew the fundamental picture was only getting worse and that the period of oversupply (Table 9.4) was likely to take hold by the end of 2008 or sometime in 2009 (see Figure 9.1). This became the prevailing view as 2008 unfolded. However, it is critical to keep in mind that

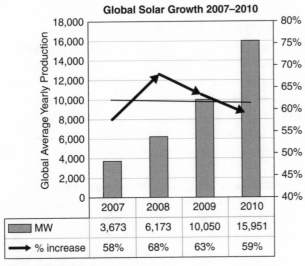

Source: AEI Estimates

Figure 9.1 Global module production estimates, 2007–2010

the savviest professional investors were well-briefed on this risk exiting 2007 and could act in their best interests ahead of the pack. The uninitiated began to understand the risk during the summer, when the subsidy picture changed dramatically for the worse with a Spanish subsidy rate reduction and cap and a reduction of subsidy rates in Germany that were worse than expected.

THE STORM BECOMES TROPICAL: SUBSIDIES CUT

As early as April the outlook for the fundamentals in the industry began to get worse, predominantly as news continued to come in from Spain and Germany threatening to bring on oversupply. By early summer advisors to both policy makers in Spain and Germany were proposing steep cuts and a possible cap in both countries.

By June 2008 sentiment for the solar stocks had turned negative again. Uncertainty about subsidy support for solar increased. Changes were coming to the German program. On June 4, 2008, the official amendment of the governmental groups that deliberated the revisions of the German renewable energy law was adopted by the leading committee on the environment. But this was not the end of the story. Over the ensuing two days the program cuts would change.

On June 5, 2008, the basic degression rate for rooftop systems greater than one MW was 12 percent in 2009, 10 percent in 2010, and 9 percent thereafter. Separately, a sliding degression rate will be applied. If installations exceed 1,500 MW in 2009, 1,700 MW in 2010, and 1,900 MW in 2011, the degression rates for *all* system sizes will be increased by 1 percent. Should total installs be less than 1,000 MW in 2009, less than 1,100 MW in 2010, or less than 1,200 MW in 2011, the degression rates for *all* system sizes will be reduced by 1 percent.

With the addition of a sliding degression rate and a concerted effort to return the benefits of the country's FIT to small system and residential operators, expectations are that the dynamics of operating in Germany must shift, as module OEMs either increase staff to maintain a high level of direct selling or, more likely, embrace distribution to handle a higher number of discrete unit sales. The change signaled an adjustment in the German market to a higher

T A B L E 9.5

German EEG: Final Results

	2009	2010	2011	2012
Outcome SPD/CDU/CSU				
<30 KW	8%	8%	9%	9%
<100 KW	8%	8%	9%	9%
<1,000 KW	10%	10%	9%	9%
>1,000 KW	25%	10%	9%	9%
FF	10%	10%	9%	9%

than expected degression rate relative to expectations set only a few days before.

On June 6, 2008, the final revisions to the German solar subsidy program, the EEG, were released (see Table 9.5). Investors were once again surprised by changes at the "eleventh hour." The large systems market degression rate was fixed at 33€ cents in 2009, down from 44€ cents—a 25 percent cut. A sliding degression rate was also set. If installations exceed 1,500 MW in 2009, 1,700 MW in 2010, and 1,900 MW in 2011, the degression rates for *all* system sizes will be increased by 1 percent. Should total installs be less than 1,000 MW in 2009, less than 1,100 MW in 2010, or less than 1,200 MW in 2011, the degression rates for *all* system sizes will be reduced by 1 percent. Despite the last-minute changes, there was no cap. Investors were relieved. Germany's role as the market of last resort was safe, at least for now. All investor eyes turned to Spain.

Very shortly after Germany resolved its subsidy program and the risk of a cap in Germany had passed, Spain leaked that it would not resolve changes to its subsidy program before late summer at the earliest. This created a level of uncertainty that professional investors realized was an advantage to them, as trading opportunities were likely to reveal themselves with every speculation about when and how the Spanish subsidy program would resolve itself before the September 2008 deadline. Counterintuitively, the savviest professional investor understood that because the resolution of the subsidy changes in Spain were likely to be negative, it created another short-term trading opportunity, provided over-supply was not developing, prices for modules were not falling, and short investors would need to run for cover ahead of quarterly reports that were most likely going to show continued strength and momentum for sales and earnings.

HURRICANES, TORNADOES, AND THE CREDIT CRISIS

The Spanish subsidy market took a turn demonstrably for the worse with a cap and steep cuts in subsidy rates as September approached and the summer heat began to give way to effects of the global credit crisis.

The first sign of the megastorm that would hit the world of finance, and simultaneously the solar industry, was the rapidly changing euro/dollar exchange rate. The solar market remains predominantly a European market. Given the rapid and volatile rise of the U.S. dollar in the FX market, professional investors grew concerned and began to analyze and judge the possible effects. On September 11, 2008, AEI took a look at these potential impacts.

Foreign Exchange Pain

In a number of cases, company disclosure on currency hedging is not adequate to perfectly estimate changes in sales. … However, we assumed an average exchange rate in 2Q08 of $1.5625/euro and a quarter-till-date average exchange rate for 3Q08 of $1.5263/euro … For the purpose of projecting our analysis into the fourth quarter, we assumed that the exchange rate on average in 4Q08 will be $1.4/euro with the further strengthening of the dollar capped at this level … SPWR has a high degree of euro exposure … Trina has 90 percent of sales in euros. … Yingli is also exposed to wide swings in currency.

—*AEI client note*

Overall market volatility, dizzying changes in foreign exchange and interest rates, and capital scarcity made it very difficult for investors to get a clear picture of the solar module oversupply scenario, but a fundamental oversupply of solar modules began to emerge in the third and fourth quarters of 2008.

T A B L E   9.6

2008 Module Companies in Production

	All Annual Module Production Capacity (MW)			
	2005	2006	2007	2008
2005	2,256			
2006	3,109	3,437		
2007	3,998	4,551	6,376	
2008	5,469	6,460	10,652	11,872
	All Annual Production Capacity Growth Rate			
	2005	2006	2007	2008
2005	base			
2006	38%	52%		
2007	29%	32%	94%	
2008	37%	42%	67%	85%

Table 9.6 presents a retrospective of the past four years of module capacity growth, focusing on year-end figures. Exiting 2008 there were approximately 11 GW of commercialized solar module production capacity and 6.5 GW of actual production, a 59 percent average industry capacity utilization rate.

The bottom line is that the oversupply situation, along with other factors, gave module buyers the leverage to hold out for, and in some cases receive, 15 to 20 percent price reductions. These price reductions accelerated into 2009 with 30 percent reductions observable across the board. Module oversupply had arrived in 2008, a year that most industry insiders believed to be just the start of great things to come. For the savvy solar stock investor who recognized the folly in the overbuilding of capacities, 2008 marked the beginning or an inevitable downturn in the solar market.

OCTOBER 2008: THE EYE PASSES OVERHEAD

There is a brief moment during a hurricane, as the eye of the storm passes directly overhead, when the sky clears and the sun shines through. It was no different in the fall of 2008 when in a surprise to everyone, the U.S. Senate leadership attached provisions of HR 6049, the Renewable Energy and Job Creation Act of 2008, to the Bush administration's bank rescue plan, which was subsequently passed and signed into law on October 3, 2008. The Renewable Energy Act renewed the solar investment tax credit, among other things, for eight years. Benefits of the law were uncapped so that the investment tax credit of 30 percent could be as large as the project. Moreover, the tax credit was extended to both individuals and corporations. Commercial and industrial companies, utilities, and residential solar customers were extended the full value of the subsidy, as the subsidy was now free from alternative minimum tax considerations.

The passage of the U.S. ITC for eight years in effect calmed markets, but only briefly, since the global economic turmoil did not end with its passage and the "bailout." It did, however, create the potential for a stable federal incentive structure. This development, although not enough to offset the pressure on solar stocks, represented a milestone in the future for solar PV throughout the world, since the duration of the support mechanism should permit certain technologies and companies to achieve the holy grail in solar: grid parity. And with that realization the eye of the storm moved off and with it took away the brief respite for those who did not recognize it for what it was. For others, not only did they realize that the solar storm would resume, but they knew that it would certainly intensify.

The Solar Storm Intensifies: The First F-5 to Strike the Industry

As 2008 drew to a close, the early signs of oversupply began to emerge, driven by the macro events in the global economy but also from within the solar industry. The tsunami of factors negatively impacting the solar business—namely, the failure of the credit markets to provide unsubsidized funding, foreign exchange moving against core markets, and pressure on the industry as supply growth continued unabated and demand fell sharply—seemed to mark the end for the solar market and stocks.

This, however, could not have been further from the truth. In the throes of the economic and financial market decline, the solar industry began to shift. The overdue consolidation necessary to rationalize the supply chain, renegotiate overpriced contracts, and redirect spending from capacity expansion to more focused technology improvements and research and development began to emerge, and with it a renewed commitment by the top players in the business to hunker down and ride out the storm.

As 2008 ended the savvy investor—the ones we often referred to in this book as having profited throughout the worst downturn ever in solar—had already begun the analysis necessary to pick the survivors and identify the changes in the solar market of the future that would itself determine the winners and losers. The first consideration in transitioning out of any storm and into better times ahead requires an assessment of just how the storm has altered the landscape. Much as it must be like emerging from the storm cellar in a hurricane ravaged community, often the shock and horror of what remains is incomprehensible to those seeking safety from the storm.

CREATIVE DESTRUCTION

As with any period of dislocation—like the one beginning in 2008 for the solar PV industry—changes to market structure occur in ways that are difficult to predict at the company level but perhaps less difficult in the aggregate. At the highest level, and certainly most obvious, traditionally weak and unsophisticated commodity segments of the supply chain will suffer the most. Wafer and module companies in the c-Si segment are likely to see the most disruption of their business models as they adjust to margins far less attractive than they were when demand was robust and supplies tight. An analysis of the solar wafer business, assuming a normalized gross margin for 16 percent and a polysilicon price of €60 per kg., results in a tremendous savings to cell makers of more than 50 percent. See Table 9.7.

Winners among the c-Si companies that emerge are likely to be companies that can capture share by reducing their supply chain costs. These are companies that successfully alter high-priced supply chain relationships, replacing them with sane contracts that are not structured like the pro-cyclical "take-or-pay" contracts with

T A B L E 9.7

Wafer Cost Reductions

Particulars	2005	2006	2007	2008	2009E
Poly contract price €/Kg	€31.73	€48.07	€57.70	€105.45	€60.14
Per gram	€0.03	€0.05	€0.06	€0.11	€0.06
g/Wp	12	11	9.7	8.5	8
Wp/per wafer	3.6	3.65	3.7	3.75	3.85
g/per wafer	43.2	40.15	35.89	31.875	30.8
€/pc wafer	€1.37	€1.93	€2.07	€3.36	€1.85
Poly cost of wafer €/Wp	€0.38	€0.53	€0.56	€0.90	€0.48
Wafer processing costs (€/Wp)	€0.32	€0.34	€0.24	€0.25	€0.21
Total wafer cost (€/Wp)	€0.70	€0.87	€0.80	€1.14	€0.69
Average selling prices (€/Wp)	€0.83	€1.82	€1.56	€1.62	€0.83
Gross margin (€/Wp)	€0.13	€0.95	€0.76	€0.47	€0.14
Gross margin %	16.19%	51.98%	48.80%	29.35%	16.36%

Source: AEI estimates

up-front deposits. Table 9-8 and Figure 9.2 provide statistical and graphic illustrations of just how powerful supply chain cost reductions are. Taking a best-of-breed c-Si module company like Yingli while holding the company's nonpolysilicon costs constant, a series of cost reductions is run through the model to illustrate the possibilities, keeping in mind that these are illustrations and not periodic estimates.

In an analysis of supply contracts that involved up-front deposits we call advance days to suppliers, a number of companies seem to have created enormous challenges to their future cost structures and competitiveness, now that polysilicon is available at declining prices. Using company filings now and in the future to work out simple formulas, investors can keep track of just how flexible and secure a company's raw material costs and volumes may be. Scouring the notes for "advances to suppliers and related parties," divided by a company's cost of goods sold, and then multiplied by 365 will remain a critical step in the investment process. As the solar storm ends, knowing the cost structure and supply chain dynamics

TABLE 9.8

The Power of Cost Reductions at Yingli

Particulars	3Q07	4Q07	1Q08	2Q08	3Q08	4Q08	1Q09E
Exchange rate —€ to $	€0.70	€0.68	€0.63	€0.63	€0.71	€0.72	€0.79
Polysilicon cost $/kg	$266.63	$277.21	$327.25	$334.61	$350.56	$311.91	$249.52
Polysilicon cost €/kg	€187.38	€189.83	€207.06	€212.07	€248.96	€224.94	€198.10
Costs in USD							
Poly cost per watt	$2.00	$2.08	$2.29	$2.31	$2.35	$2.00	$1.57
Non poly cost per watt	$0.82	$0.80	$0.80	$0.80	$0.80	$0.80	$0.80
Total cost per watt	**$2.82**	**$2.88**	**$3.09**	**$3.11**	**$3.15**	**$2.80**	**$2.37**
Costs in EUR							
Poly cost per watt	€1.41	€1.42	€1.45	€1.46	€1.67	€1.44	€1.25
Non poly cost per watt	€0.57	€0.55	€0.51	€0.51	€0.57	€0.58	€0.64
Total cost C/W	**€1.98**	**€1.97**	**€1.96**	**€1.97**	**€2.24**	**€2.02**	**€1.88**
Silicon usage (gms/w)	7.50	7.50	7.00	6.90	6.70	6.40	6.30

271

TABLE 9.8

The Power of Cost Reductions at Yingli (Continued)

Particulars	2Q09E	3Q09E	4Q09E	1Q10E	2Q10E	3Q10E	4Q10E
Exchange rate —€ to $	€0.80	€0.80	€0.80	€0.80	€0.80	€0.80	€0.81
Polysilicon cost $/kg	$199.62	$159.70	$127.76	$102.21	$81.76	$65.41	$52.33
Polysilicon cost €/kg	€158.80	€127.29	€102.04	€81.79	€65.56	€52.56	€42.13
Costs in USD							
Poly cost per watt	$1.24	$0.97	$0.77	$0.60	$0.47	$0.37	$0.29
Non poly cost per watt	$0.80	$0.80	$0.80	$0.80	$0.80	$0.80	$0.80
Total cost per watt	**$2.04**	**$1.77**	**$1.57**	**$1.40**	**$1.27**	**$1.17**	**$1.09**
Costs in EUR							
Poly cost per watt	€0.98	€0.78	€0.61	€0.48	€0.38	€0.30	€0.24
Non poly cost per watt	€0.64	€0.64	€0.64	€0.64	€0.64	€0.64	€0.64
Total cost C/W	**€1.62**	**€1.41**	**€1.25**	**€1.12**	**€1.02**	**€0.94**	**€0.88**
Silicon usage (gms/w)	6.20	6.10	6.00	5.90	5.80	5.70	5.60

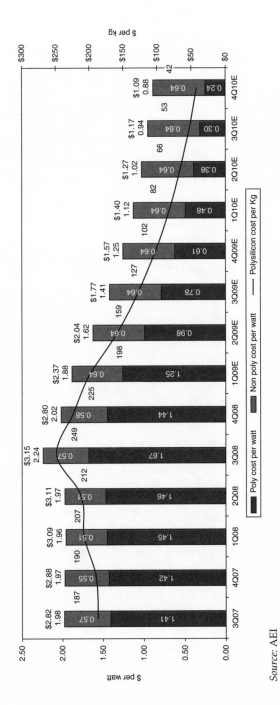

Source: AEI

Figure 9.2 Graphic illustration of the power of cost reductions at Yingli

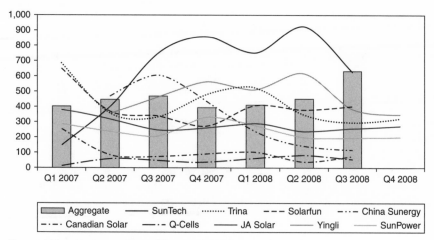

Source: AEI estimates, company filings

Figure 9.3 Solar advance to suppliers days

going on at a solar company and across the industry will be required homework for the solar stock investor. See Figure 9.3.

Polysilicon companies, although having complex processes and high capital costs, are also likely to revisit very low or negative margins as supplies exceed demand in the marketplace. Figure 9.4 illustrates how the polysilicon supply is forecast to outstrip demand.

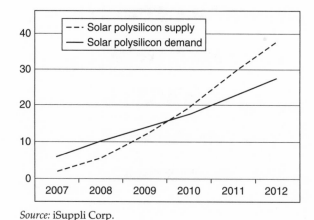

Source: iSuppli Corp.

Figure 9.4 Polysilicon supply vs. demand projections

A key for investors in polysilicon companies going forward will be to estimate the industry's ability to meet market growth, recognizing commodity dynamics and applying commodity industry investment techniques.

The lesson revealed during the 2005–2007 bull market for polysilicon is to pay a peak multiple at the cyclical bottom for polysilicon selling prices and to compress the multiple one is willing to pay on the way up so the peak multiple one is willing to pay is often when the polysilicon selling price is at its lowest. The polysilicon price history and forecast is graphically summarized in Figure 9.5.

An observation about estimating costs in solar that we made in earlier chapters is the ailed notion of "backstop markets" that may have contributed to overbuilding and the resulting fallout. Estimating solar module demand, assuming the presence of a belief in backstop market ideology, may lead to very large capacity-driven values. During the peak of the boom years for solar, one such estimator joined forces with *Photon* magazine, a very widely read industry periodical published in many languages, and as a result the world was flooded with extremely high estimates for industry

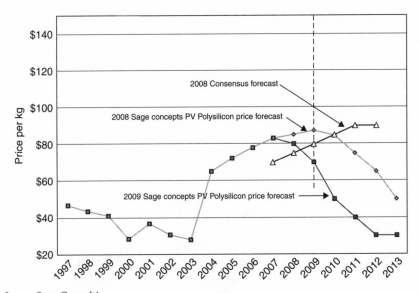

Source: Sage Consulting

Figure 9.5 Average polysilicon price history and forecast

growth, margins, and profits. During this period the ideology that backstop markets had the capacity to absorb module supplies given the demand-elasticity assumed going forward seemed to create an aura around industry executives that I witnessed when working as a partner at a hedge fund on Wall Street. Although the very notion of backstop markets has been clearly revealed as unreliable, the estimates of certain estimators has not yet relented fully.

Suntech Power is a company that engaged in raw material contracts on a take-or-pay basis that seemed to follow the bull-market mode—"contract for scarce supplies of polysilicon, raise capital based on having what others had little of (read 'polysilicon'), and expand." Table 9.9 illustrates a series of different deals and investments one company, Suntech, entered into, apparently following a formula of locking up supplies in order to expand at a breakneck speed.

Rationalizing the supply chain is likely to occur from 2009 to 2011 as margins shift and the weakest companies retreat from the industry or are rationalized into fewer companies. The foundational

TABLE 9.9

Suntech Supply Deals and Investments (2007–2008)

Supplier	Time Frame	Volume
MEMC	2007–2016	$5–6 billion
REC	2007–2011	$180 million
Sunlight	2007–2011	$366–$670 million
Nitol Solar	2009–2015	Unspecified
Asia Sil	2H08–2015	$1.5 billion
Hoku	2009–2018	$678 million
Shunda	2008–2020	7,000 MW
DCC	2009–2016	$631 million
Wacker	Unspecified	220 MW
PV CRY	2008–2013	260 MW
GCL Wafer	2008–2012	9.4GW + 1.1GW
DCC #2	2010–2016	$750 million
ReneSola	2008–2016	1.5 GW
Glory	Four years	Unspecified

premise likely to guide the market during the reorganization phase will be cost reduction. The solar industry, on average, must reduce costs 50 percent to achieve grid parity in a broad geographic sense. Until grid parity is met, the cycles of the industry will remain captive to growth in global subsidy programs.

Will rationalization alone lead to grid parity for the industry? Unfortunately, it will not. However, the strongest companies with technology, market access, and financial superiority are likely to benefit from consolidation. The solar investor may be surprised to see an astonishingly high number of companies go out of business as the strongest companies throughout the supply chain combine and direct their consolidated resources to cost-saving programs and research. Thin film companies, however, may observe a different dynamic.

Thin film companies will find themselves in the same declining selling price environment as the c-Si based solar companies, but they may seek to use their already advantaged low price to absorb even lower prices by accepting a temporary hit to their margins and cutting costs as quickly as possible while continuing to increase manufacturing capacity. Should such a dynamic occur in the industry, the risk to c-Si solar companies may be an accelerated market share loss to the thin film solar companies and the slowing of production over a protracted period. In a market that gains share for thin film, the investor would likely witness consistent outperformance of thin film solar stocks on a relative basis compared to the c-Si solar group. Investment opportunities may also arise as c-Si solar companies diversify technology risk and acquire smaller thin film companies.

Although they have the advantage of a lower systems cost, thin film companies must also accelerate their cost reduction programs to restore margins quickly and widen their cost advantage at the systems price level. They already benefit from a more integrated supply chain, stemming from a consolidated manufacturing process, compared to c-Si solar companies. Vertical integration, therefore, does not hold the same benefits for the thin film companies as it does for the c-Si group. However, thin film modules have the potential to grow in size, and therefore thin film companies have opportunities to improve in energy density to achieve more power output per area, as well as integrate balance of systems costs to ameliorate a technology disadvantage when compared to c-Si solar PV. Thin film companies must improve efficiency and simultaneously integrate balance of system components (inverters, charge controllers, etc.). They may accelerate their cost reduction programs and possibly acquire companies with advanced balance of system component technology, while maintaining their capacity ramp programs.

The shortages throughout the supply chain during the 2005-2008 bull market for solar impacted a wide variety of suppliers. DuPont could not make enough of its Tedlar product, a wide array of equipment companies had year-long backlogs, and a shortage of solar glass created bottlenecks as important as the polysilicon shortage during the same period. The key to identifying investment opportunities in the new postcrisis era for solar PV will also depend on understanding which suppliers did not overbuild capacity during the boom, and if these suppliers offer pure-play investment opportunities. Other investment opportunities will be derivatives such as suppliers acquiring device companies to capture future growth and promote cost savings, perhaps through integrated facilities.

Solar glass comes to mind. Currently, glass companies serving the architectural, automotive, and LCD markets have tremendous growth opportunities tied to solar PV. Glass in the solar PV industry is expensive mainly because of a small batch manufacturing process dictated by the smaller scale of today's solar PV industry, glass thicknesses that are three times those used in the LCD market, and transportation costs associated with making the glass far from the solar fab. For the thin film solar company with glass-on-glass large scale modules, there are potentially profound cost saving opportunities to colocate very large scale solar manufacturing plants adjacent to interconnected dedicated glass fabricating facilities. Will companies like PPG Industries acquire a solar thin film company?

U.S. UTILITY MARKET

The passage of the U.S. ITC created renewed hope that a solar PV utility market might emerge, thereby lifting the demand for solar and supporting growth. Although the growth of the utility market is advantageous, one should not expect the utility market to burst onto the scene in 2009. It will take the utility companies longer to ramp. The U.S. utility market will more likely gain traction beginning in 2010, when the lion's share of the volume from project contracts signed in 2008 are scheduled to be installed. To get a picture of just how much demand could be stimulated by the emerging U.S. utility market, investors will take note of the tax savings available to utility companies that choose solar PV for a more substantial part of their renewable portfolio standards plan. Table 9.10 was presented in November 2008 at the SunPower Corporation analyst day. SunPower claims the utilities listed in the chart will find the tax advantages too attractive to pass up.

Although the emergence of the U.S. utility market is as attractive to the solar stock investor as to solar companies, it is not without the potential for risk. A barrier to deep and volume-driven commodity markets for solar PV in the United States has been in part due to the lack of significant involvement on the part of utilities to play a role in the development of the industry (that is, systems ownership). The emergence of the U.S. utility market presumes this limiting factor goes away without any significant change to business models, both at the utility itself and at any part throughout the solar PV supply chain.

Today's solar PV business models center on user-owned and third-party-owned solar PV systems financed by owners, power purchase agreements, and tax equity pools of capital. However, to create large markets for solar PV in the United States, the passive role of utilities must change to accommodate a changing paradigm brought on by the disruption of their existing business models and to best capitalize on solar PV's unique benefits. It is not sufficient to view the emerging utility market as the opening up of large markets for solar PV in the United States solely because of the passage of the U.S. ITC and a few utility contracts.

The utility market opportunity should be considered one capable of achieving double-digit penetration of utility peak-load power generation. Penetration rates in excess of single digits will be driven by local, regional, and national electric grid infrastructure concerns—not tax planning. Utilities, therefore, must engage in capital

TABLE 9.10

Utilities Add Substantial Tax Capacity

Estimated Taxes ($MM)	FY 08	FY 09	FY 10
Exelon	$1,954	$2,030	$2,034
Duke	1,106	1,627	1,495
Southern	1,087	1,282	1,396
FirstEnergy	959	1,182	1,234
Public Service Enterprise	934	1,020	1,171
AES	764	990	346
Edison Intl	680	918	1,044
Entergy	780	913	991
Dominion	607	897	1,033
PG&E	675	824	889
Consolidated Edison	709	713	823
FPL	524	690	934
Rest of top-30 utilities	4,348	6,547	8,108
Total estimated taxes	**$16,126**	**$19,634**	**$21,497**
Estimated solar ITC capacity			
Tax equity allocated to solar (est. 10%)	**$1.6 B**	**$2.0 B**	**$2.2 B**
Solar investment capacity w/ 30% ITC	**$5.4 B**	**$6.5 B**	**$7.2 B**
Solar project MW capacity (@ $6/W)	**896 MW**	**1,091 MW**	**1,194 MW**

Source: SunPower

planning on a broad scale, including changes to traditional capacity additions, transmission, and quality of service considerations burdened by concerns for business model realities, including revenue dilution should nontraditional solar PV power generators enter the scene. The future of the U.S. utility market, therefore, becomes one of potential risk for those solar PV companies that today are potentially planning to morph their business models into solar PV electricity generation and distribution companies. The third-party model for integrated device manufacture with self-directed project development could lead to commodity gross margins far below the gross margin expectations of today.

Furthermore, as utilities plan to achieve deeper penetration of solar PV into their mix, their corporate strategies could easily adapt to accommodate system sales, marketing and financing, and maintenance, potentially destroying the value in nonutility solar PV downstream investments. Although SunPower is a preferred solar system provider to utilities today, its investment in PowerLight, the systems arm of the company that constitutes more than 50 percent of sales, could see its business model jeopardized by a truly integrated utility market of the future.

EMERGENCE OF SOLAR MASTER LIMITED PARTNERSHIPS

As the U.S. market becomes the global epicenter of future solar PV system installs, the investor may look to possible changes in the tax code or investment opportunities. As mentioned earlier in this chapter and elsewhere throughout the book, the U.S. ITC passed late in 2008 as part of the bailout will have stimulative effects on the growth rate for solar PV. Unfortunately, because the macroeconomic picture for the remainder of the decade has been downgraded, the use of "tax credits" must also be downgraded, thereby slowing the capital formation to support new PV systems growth.

This, however, has not gone unnoticed by the Solar Energy Industries Association leadership which, immediately after passage of the U.S. ITC, began a concerted effort to promote positive changes to the tax code to allow "tax refunds" or "tax credits"— tax refunds in a slow economy would have great appeal where tax credit demand was stymied by poor returns associated with slow or no growth economic conditions.

Second, the Solar Energy Industries Association leadership is also seeking changes to the tax code to permit master limited partnerships to facilitate the formation of large pools of capital to fund solar PV growth. The mutual fund and retail investor who sought safety during the stock market route of 2008 took refuge in natural gas pipeline master limited partnerships and did very well. Should the tax code be amended to permit master limited partnerships to be used as investment vehicles funneling capital to solar PV (that is, extending the 30 percent refund/credit to the MLP), a new and very exciting investment opportunity will arise.

AFTER THE STORM: SOLAR PV BUSINESS
MODELS OF THE FUTURE

All storms, whether hurricanes or tornadoes, eventually end, and in cases like the storm visited on the solar industry, the solar storm will eventually end as well. What will be left in its wake will be a rebuilding effort, a drive toward modified or new business models and required innovations in distribution and solar system and project finance. These changes must all contribute to an overall lower cost of solar electricity production and the lowest cost of capital for the industry. However, as the solar PV industry is truly in its infancy, a series of business models will likely be tried out and fail until the goals of achieving the lowest production cost on a levelized cost of energy and the lowest cost of capital are achieved. Current business models are predicated on best-of-breed, vertical integration, and emerging solar utility business models. Each business model is likely to offer investors opportunity to make—or, more likely, lose—money. The savvy investor entering the light of a new day after the solar storm will select those companies that can flex their supply chains, achieve lower costs, and manage their profitability where others cannot. Such an investor will also be on the lookout for companies that can adapt their business models to retain ownership of the system, selling energy as a new source of revenue. Although these business model changes may sound attractive, they come with the potential for significant investor risk. Should SunPower innovate its business model in this manner and not preserve continuity in the investor base, or introduce undesired financing risk, or fail to mitigate the liability stemming from the business of becoming a solar utility, the company's stock could suffer. Of course, SunPower is not the only company to consider the path of the emerging solar utility business model. Other high-profile companies like First Solar have also openly taken steps in this direction.

As we move into the spring of 2009, we remain in the throes of the global credit crisis, and the unprecedented storm that has beset the solar industry isn't over—yet. We are, in fact, in "the worst of times," as Charles Dickens famously wrote as he began his great novel, *A Tale of Two Cities* (1859). As in Dickens's book, the hope of a rebirth for the solar stocks after the credit crisis and the foreshadowing of better times ahead are not easy to puzzle out.

But better times are ahead, and despite the investment risks, solar power is perhaps the single largest investment opportunity ever to come along. The foundations of the opportunity do not change, no matter what financial crisis befalls us. Climate change and the inexorably higher costs of carbon-intensive energies remain. Solar energy and market development have a clear path to produce clean energy for less than carbon-intensive fuel sources of energy. Solar energy fuel is abundant and free. It is a key step in enabling sustainable economic development and generating geopolitical political accord. Last but by no means least, to this author and millions of others, solar energy provides an opportunity to have energy independence and security for all. Depending on how wisely we invest, we stand to reap rich returns for generations to come making money from investing in solar stocks.

INDEX

ABOUT THE AUTHOR

Joseph Berwind, founder of Alternative Energy Investing™ (AEI), has ten years of experience that includes portfolio management and research in energy-technology, renewable-fuel, advanced materials, and technology equity research. AEI's clients include many of the top-ten institutional holders in the solar stock industry. In his role as managing partner, Berwind is responsible for the firm's day-to-day solar stock research mandate, and additional energy-technology and renewable-fuel sector research.